Landscape ecology and
geographic information systems

18

WITHDRAWN

9

1

Landscape ecology and geographic information systems

Edited by

Roy Haines-Young   **David R. Green**   **Steven Cousins**

Taylor & Francis
London ● New York ● Philadelphia
1993

UK   Taylor & Francis Ltd, 4 John St, London WC1N 2ET

USA   Taylor & Francis Inc., 1900 Frost Road, Suite 101, Bristol PA 19007

British Library Cataloguing in Publication Data

A catalogue record for this book is available from the British Library

ISBN 0 7484 0002 8 (cloth)
ISBN 0 7484 0252 7 (paper)

Library of Congress Cataloging in Publication Data are available

Cover design by Amanda Barragry derived from original artwork 'LANDSCAPE' by E. G. Green

Typeset by Santype International Limited, Netherhampton Road, Salisbury, Wilts SP2 8PS

Printed in Great Britain by Burgess Science Press, Basingstoke on paper which has a specified pH value on final paper manufacture of not less than 7·5 and is therefore 'acid free'.

Contents

Contributors

R. Aspinall
Division of Land Use Research, Macaulay Land Use Research Institute, Craigiebuckler, Aberdeen AB9 2QJ, UK

P. B. Bridgewater
Australian National Parks and Wildlife Service Head Office, GPO Box 636, Canberra, ACT 2601, Construction House, 217 North Bourne Avenue, Turner ACT 2601, Australia

S. Cousins
International Ecotechnology Research Centre, Cranfield Institute of Technology, Cranfield, Bedford MK43 0AL, UK

R. Cummins
Institute of Terrestrial Ecology, Hill of Brathens, Banchory, UK

R. Dunn
Department of Geography, University of Bristol, University Road, Bristol BS8 1SS, UK

P. J. Edwards
Geodata Institute, University of Southampton, Southampton SO9 5NH, UK

G. M. Foody
Department of Geography, University College of Swansea, Singleton Park, Swansea SA2 8PP, UK

P. A. Furley
Department of Geography, University of Edinburgh, Edinburgh EH8 9XP, UK

R. Goossens
Laboratory for Regional Geography and Landscape Science, State University of Ghent, Krijgslaan 281 (S8-A1), B-9000, Ghent, Belgium

D. R. Green
Centre for Remote Sensing and Mapping Science, Department of Geography, University of Aberdeen, Aberdeen, UK

G. H. Griffiths
18 Cobden Crescent, Oxford, OX1 4LJ, UK

H. Gulinck
Katholieke Universiteit Leuven, Faculty of Agricultural Science, Kardinaal Mercierlaan 92, B-3030 Leuven, Belgium

A. M. Gurnell
Geodata Institute, University of Southampton, Southampton SO9 5NH, UK

R. Haines-Young
Department of Geography, University of Nottingham, Nottingham, UK

E. D'Haluin
Laboratory for Regional Geography and Landscape Science, State University of Ghent, Krijgslaan 281 (S8-A1), B-9000 Ghent, Belgium

A. R. Harrison
Department of Geography, University of Bristol, University Road, Bristol BS8 1SS, UK

R. G. Healey
Department of Geography, University of Edinburgh, Edinburgh EH8 9XP, UK

I. Heywood
Department of Geography, The University, Salford M5 4WT, UK

C. T. Hill
Geodata Institute, University of Southampton, Southampton SO9 5NH, UK

P. Janssens
Katholieke Universiteit Leuven, Faculty of Agricultural Science, Kardinaal Mercierlaan 92, B-3030, Leuven, Belgium

C. A. Johnston
Natural Resources GIS Laboratory (NRGIS), Natural Resources Research Institute, University of Minnesota, 5031 Miller Trunk Highway, Duluth, MN 55811, USA

J. Kolejka
Department of Environmental Science, Masaryk University, Kotlarska, 2, 61137 BRNO, Czechoslovakia

G. Larnoe
Laboratory for Regional Geography and Landscape Science, State University of Ghent, Krijgslaan 281 (S8-A1), B-9000, Ghent, Belgium

C. J. Lavers
Department of Geography, University of Nottingham, Nottingham, UK

R. A. MacMillan
Alberta Research Council, PO Box 8330, Postal Station F, Edmonton, Alberta T6H 5X2, Canada

J. Miles
Institute of Terrestrial Ecology, Hill of Brathens, Banchory, UK

R. J. Naiman
Centre for Streamside Studies, AR-10, University of Washington, Seattle, WA 98195, USA

T. Onega
Laboratory for Regional Geography and Landscape Science, State University of Ghent, Krijgslaan 281 (S8-A1), B-9000, Ghent, Belgium

J. Pastor
Natural Resources GIS Laboratory (NRGIS), Natural Resources Research Institute, University of Minnesota, 5031 Miller Trunk Highway, Duluth, MN 55811, USA

F. Perez-Trejo
International Ecotechnology Research Centre, Cranfield Institute of Technology, Cranfield, Bedfordshire MK43 0AL, UK

J. R. Petch
Department of Geography, University of Salford, Salford M5 4WT, UK

D. A. Stow
Department of Geography and Systems Ecology Research Group, San Diego State University, San Diego, CA 92182, USA

N. Veitch
ITE, Monkswood Experimental Station, Abbots Ripton, Huntingdon, PE17 2LS, UK

O. Walpot
Katholiek Universiteit Leuven, Faculty of Agricultural Science, Kardinaal Mercierlaan 92, B-3030 Leuven, Belgium

T. F. Wood
MVA Systematica, MVA House, Victoria Way, Woking, Surrey GU21 1DD, UK

R. Wright
Centre for Remote Sensing and Mapping Science, Department of Geography, University of Aberdeen, Aberdeen, UK

Acknowledgement

Dr D. R. Green wishes to acknowledge the assistance given by the Cartographic Office, Department of Geography, University of Aberdeen in re-drawing figures within chapters 6, 9, 16 and 19

PART I

Introduction

1

Landscape ecology and geographical information systems

R. Haines-Young, D. R. Green and S. Cousins

The landscape perspective

An appreciation of landscape is something we share with our earliest ancestors. For them, an understanding of the resources associated with the land meant the difference between the prosperity or collapse of their communities. For current generations, particularly those in the developed world, our relationships with the landscape are more subtle—but just as important culturally and economically.

Modern technologies have distanced us from many of the rigours associated with gaining a living from the land. So much so, that many have come to regard the landscape simply as a backdrop to our daily lives rather than a resource which needs to be nurtured and managed. Technology may have blurred the links which we have with landscape, but the link is as important now as it ever was in the past.

In both the developed and developing world we are engaged in massive transformations of the natural and semi-natural vegetation cover of the Earth. In other areas traditional landscape patterns, which have been largely stable or only slowly changing for may generations, are now being rapidly altered. The consequences of such change are unknown, but some have argued that they may be profound.

In the tropics, the loss of rain forests is undermining a major component of the genetic resource base of the biosphere. The destruction of the tree cover may also have significant effects on a range of climatic parameters, through the release of carbon dioxide and water (Woodwell *et al.*, 1984) or changes in surface albedo (Verstraete *et al.*, 1990). Elsewhere in tropical and subtropical areas, where the wood is a major source of domestic fuel, the pressure of human populations on semi-arid scrub communities may be a major contribution to the process of desertification and the collapse of essential 'life support systems' in these areas.

In Europe, North America and other industrialized parts of the world, effects on the landscape are more indirect, but equally significant. The generation of acid rain and its consequences for forests and lakes, for example, is as serious a problem as anything encountered in the developing world. At a time when we are seeking to minimize forest loss caused by clearance, increased pollution loads in industrialized areas threaten the integrity of many forested landscapes. The consequences are not just of local and regional significance, but may also impact upon the global system.

As a species we cannot help but modify our environment. Again this is a characteristic we share, not just with our ancestors, but also with all other organisms. The problem we now face is that of the speed and scale at which these changes are occurring, which may mean that biological, social and economic systems will find it difficult to adjust.

In seeking to understand the significance of environmental change, we are increasingly being forced to take an integrated perspective. Environmental problems rarely respect conventional subject boundaries, and their solution requires both an understanding of the physical and ecological aspects of environmental systems and the way in which they interact with economic, social and political factors. The discipline of landscape ecology provides part of this integrated view.

Landscape, according to Vink (1983), is the sphere in which a range of processes are active. Landscape ecology aims to focus on the way in which these processes interact, and provide a framework in which human impact on the environment can be understood. From understanding, actions and suitable management strategies can be developed.

The first use of the term 'landscape ecology' is generally credited to the German geographer Carl Troll who saw it as the union of geography and ecology. Subsequent definitions of the subject have been many and varied. Thus, while Vink (1983) sums up many of the more traditional views with his assertion that the study of landscape

> ... is the study of the relationships between phenomena and processes in the landscape or geosphere including the communities of plants, animals and man. (Vink, 1983: 2)

Forman and Godron (1986) give a more technical definition of landscape ecology as:

> ... the study of the structure, function and change in a heterogeneous land area composed of interacting ecosystems. (Forman and Godron, 1986: 595)

Subject definitions are convenient in introductory texts, but need not detain us too long here. The most important thing is not what we take a subject to include, but whether the concepts and theories a discipline develops help us understand and deal with the world around us.

Geographical information systems

One of the difficulties which we faced as landscape ecologists was that, until quite recently, the analytical tools available did not match the scale of questions we needed to ask about landscapes. At the regional scale consistent data about the Earth's surface and its cover were difficult, time consuming, and expensive to collect. Large data volumes also made processing complex and integration with other data difficult. Fortunately, with the availability of computer-based systems for handling geographical or spatial data, called 'geographical information systems' (GIS), many of these difficulties are beginning to be overcome.

In this book we use the term 'GIS' in a very general way to describe any computer-based system for the input, storage, analysis and display of spatial information. We also include within the scope of this definition systems designed primarily to capture spatial information and also to process it, namely remote sensing systems. Over the last decade or so the development of ideas in landscape ecology has increased significantly. The stimulus for some of this work has been better

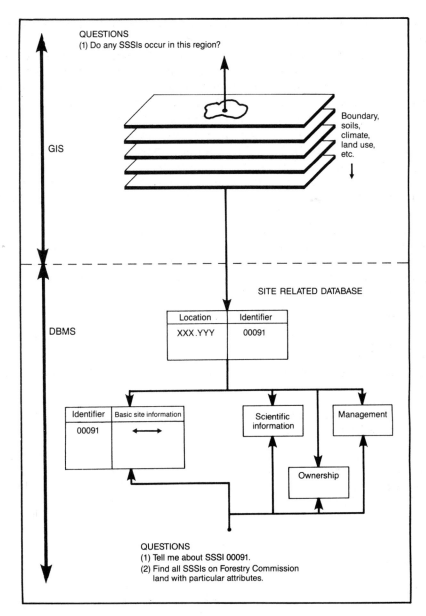

Figure 1.1. Some basic elements of a geographical information system (GIS). In the example shown, users may query the information relating to sites of special scientific interest (SSSIs) either through the map data or through the attribute information held in a database management system (DBMS).

access to these computer-based systems for handling spatial information. This volume highlights some of the achievements that have been made in developing the tools and concepts of landscape ecology. The subject is now being driven by the International Geosphere Biosphere Program (IGBP) and global/regional issues, e.g. desertification.

One way of visualizing what a GIS can do is to think of them as being able to handle many layers of map information relating to an area (Figure 1.1). Each layer describes a different aspect of its geography. One layer might hold data on geology, another on soils. Subsequent layers might include data on land cover in the area, species distributions, or the socio-economic characteristics of the human population in the area. The power of GIS lies in the fact that data from any combination of these layers might be used to solve a particular problem. Furthermore, as problems change, the data can be processed in different ways to address different issues in a highly flexible way. Although the ability to handle spatial information in the form of maps is important, GIS can also hold non-spatial attribute information which can be associated with the various map features in a database management system of some kind. These data can also be used to access the map information. Thus, in the example shown in Figure 1.1, a query about the location of sites of special scientific interest (SSSI) might be made starting either from the location on one of the map layers or by means of text data relating to the ownership or management status of the site.

A fuller account of the technical background to GIS, with particular reference to the landscape ecologist, can be found in Johnson (1990). Burrough (1986) and Maguire *et al.* (1991) provide more extensive treatments of the topic. The present volume also provides a number of chapters which provide examples of applications that use the technology. This material illustrates that, despite the versatility of GIS as powerful analytical tools, the general problem of data availability is a major barrier to their wider use.

The range of data held in a GIS depends upon the intended use of the system. Users largely purchase a 'shell' which they populate with data relevant to their needs. Some of the users will have to convert data into machine-readable form themselves, through the process of digitization. Some data may come from third-party sources, in the form of digital-map information.

In recent years a major 'third-party' source of digital information for landscape ecologists has been the data provided by the various types of air- and space-borne remote sensing systems. Remote sensing is simply the acquisition of information about the land, sea and atmosphere by sensors located at some distance from the target of study. The sensors usually depend on the spectral properties of the target as the basis of the measurements they make. In general, these systems record information in digital format, which can be processed back on Earth to reveal information about a whole range of surface characteristics of interest to the ecologist (Wickland, 1989; Hobbs and Mooney, 1990; Haines-Young, in press). More general reviews of the field are provided by Colwell (1985) and Mather (1987). This volume also provides a number of case studies which illustrate something of the versatility of these systems.

The challenge of the new technologies

As landscape ecologists, the questions we need to ask about the landscape and the human impact upon it are complex and highly demanding of intellectual frameworks. The growth of the discipline has in recent years been stimulated by access to the new technologies for handling spatial information, which may help us to overcome some of the practical difficulties we face. But investment in technology is only worthwhile if it allows us to solve outstanding scientific problems or to look at the world in new and more perceptive ways. For those of us who work with spatial information systems, this is the challenge. This book is intended as a contribution to the debate.

The material collected together in this volume is split into four sections. Part II provides an overview of the use of remote sensing and GIS in landscape ecology. Part III looks at some conceptual issues arising out of the use of the new technology, and Part IV considers some analytical techniques and technical issues. The final and most substantial section of this book provides a series of papers describing how spatial information systems are being used to tackle problems in landscape ecology.

The core of the material presented here arose from a workshop held in the Department of Geography at the University of Nottingham, in January 1989. This material was extended by inviting several additional paper Landscape ecology is a very broad and rapidly developing discipline. This volume emphasizes the importance of water flow in structuring the abiotic environment and the importance of the movable part of the biotic system, animals, in relation to the landscape. The key to plant distributions lies with both these influences, because plants are important water transporters and evaporators as well as being the food base for animal populations. The interaction of water flow, plants and animals is a two-way process which forms and is formed by the landscape. Specific interactions between these elements of the landscape are explored in this book which illustrates the richness of the emerging relationships between the theory of landscape ecology, data capture and GIS.

References

Burrough, P., 1986, *Principles of Geographical Information Systems*, 2nd Edn, Oxford: Oxford University Press.

Colwell, R. N., 1985, *Manual of Remote Sensing*, 2nd Edn, Falls Church, Virginia: American Society of Photogrammetry.

Forman, R. T. and Godron, M., 1986, *Landscape Ecology*, New York: Wiley.

Haines-Young, R. H., in press, Remote sensing, GIS and the assessment of environmental change, in Roberts, N. (Ed.), *Global Environmental Change*, Oxford: Blackwell.

Hobbs, R. J. and Mooney, H. A., 1990, *Remote Sensing of Biosphere Functioning (Ecological Studies No. 79)*, New York: Springer-Verlag.

Johnson, L. B., 1990, Analysing spatial and temporal phenomena using geographical information systems—a review of ecological applications, *Landscape Ecology*, **4**, 31–44.

Maguire, D. J., Goodchild, M. F. and Rhind, D. W., 1991, *Geographical Information Systems*, 2 Vols, London: Longman Scientific and Technical.

Mather, P. M., 1987, *Computer Processing of Remotely-sensed Images*, London: Wiley.

Verstraete, M. M., Belward, A. S. and Kennedy, P. J., 1990, The Institute for Remote Sensing Applications contribution to the global change research programme. 'Remote sensing and global change', in Coulson, M. G. (Ed.), *Proc. 16th Annual Conference of the Remote Sensing Society*, pp. iv–xii, Swansea: University College.

Vink, A. P. A., 1983, in Davidson, D. A. (Ed.), *Landscape Ecology and Land Use*, London: Longman.

Wickland, D. E., 1989, Future directions for remote sensing in terrestrial ecological research, in Asrar, G. (Ed.), *Theory and Applications of Optical Remote Sensing*, pp. 691–724, New York: Wiley.

Woodwell, G. M. (Ed.), 1984, *The Role of Terrestrial Vegetation in the Global Carbon Cycle: Measurement by Remote Sensing*, New York: Wiley.

PART II

Overviews

2

The role of geographic information systems for landscape ecological studies

D. A. Stow

Introduction

Assessments of patterns of ecosystem structure and function are based on spatially distributed ecological data, which are necessarily recorded at a variety of spatial and temporal scales. These data, particularly those derived from remotely sensed images, may be more efficiently stored and more effectively analysed using a geographic information system (GIS) (Risser and Treworgy, 1985).

The objective of the present chapter is to specify the role that GIS has or soon will be playing in the study of landscape ecological and other spatial biophysical processes. Examples are primarily chosen from more arid landscapes (including Mediterranean, arctic tundra and desert ecosystems), because of the author's interest and recent work in such landscapes.

The use of GIS technology in ecosystems research is a recent phenomenon, which has rapidly become part of the mainstream in research such as the US National Science Foundation's Long Term Ecological Research (LTER) Program (Swanson and Franklin, 1988). However, the literature is currently lacking in research that involves the use of GIS for scientific studies of ecosystems on the landscape scale or larger (Davis and Dozier, 1988). Most studies reporting the development and usage of GIS have been oriented towards resource-management applications rather than ecological studies (Kessel and Cattelino, 1978; Yool *et al.*, 1985).

Of particular importance to landscape ecology is the need for developing GISs that handle ecological data of a variety of scales in a hierarchical fashion. Such systems should support the following functions:

1. provide a database structure for efficiently storing and managing ecosystems data over large regions;
2. enable aggregation and disaggregation of data between regional, landscape and plot scales;
3. assist in the location of study plots and/or ecologically sensitive areas;
4. support spatial statistical analysis of ecological distributions;
5. improve remote-sensing information-extraction capabilities; and
6. provide input data/parameters for ecosystem modelling.

These GIS-related functions are elaborated on further in the context of the ecological and geographical research literature, again with emphasis on semi-arid and arid ecosystems.

Database structure

It is apparent that with the myriad of field and remotely sensed data which must be assimilated in landscape ecology analyses, it is necessary to establish a computerized database. Risser and Treworgy (1985) provide some excellent reasons for developing such a database for ecological science and management, most of which are concerned with the need to integrate data of numerous and complex forms. They also outline many of the critical issues in the development, usage and management of an ecological database. At the end of their overview paper, Risser and Treworgy (1985) state that a GIS is a logical choice for structuring an ecological database.

Storing, retrieving and analysing ecological data by geographic coordinates and by using spatial data structures are powerful bases for establishing a GIS for multi-scale studies of ecosystems (Marble *et al.*, 1984). Wells and McKinsey (1990) have shown that a raster GIS database structure is an effective means for making fire-management decisions in southern California parklands. The GIS structure was effective for both locational siting analyses and for spatial modelling purposes. Raster-coded GIS are also inherently compatible with digital satellite image data (Jensen, 1986). The major advantage of raster structures for landscape ecology applications is their representation of continuous or surface-type data, such as elevation, surface temperature or biomass.

Vector data structures have also been shown to be effective for ecological/resource management studies, and provide a more efficient structure for data storage (Maffini, 1987). While spatial overlay modelling may be more straightforward with raster structures, the results of recent attempts at integrating GIS with spatially explicit models of landscape processes (e.g. surface run-off models) suggests that vector coding provides a powerful structure for achieving this integration (Silfer *et al.*, 1986; Haber and Schaller, 1988).

Hierarchical format

The range of spatial and temporal scales of ecosystem processes and the desire to assess ecological structure at the landscape and regional scales, gives rise to the need to integrate hierarchical aspects of ecological theory and database formats. GIS is a database tool that can be used effectively to handle environmental data of a variety of scales (Risser, 1986).

The hierarchical nature of ecosystem processes and structure is justifiably receiving more attention (Allen and Starr, 1982; Delcourt *et al.*, 1983; Steele, 1985), as the interdisciplinary field of landscape ecology continues to develop. As hierarchical theory progresses, ecologists should have a stronger basis for moving between different scales of ecosystem process. However, the infancy of ecological theory in terms of linkages between scales (Risser, 1986) means that it is still necessary to explore empirical aggregation/disaggregation methods for assessing scales of ecosystem structure and function. Pertinent to this theme, Allen and Hoekstra (1984) have eloquently explained the importance of choosing appropriate spatial scales of measurement, when there is a requirement to aggregate or disaggregate ecosystem processes between scales. One can also infer from their treatise that the data structure used to store and retrieve environmental data in a GIS imposes an artificial pattern that will influence how successfully the ecosystem scientist can analytically move between scales.

A multi-scale GIS can provide a mechanism for performing empirical evaluations of scale variations of ecosystem function and structure (Walker *et al.*, 1990). Such a GIS should build on principles of multi-stage sampling (Langley, 1969). One approach to creating a multi-scale GIS is through the use of quad-tree structures (Chen and Peuquet, 1985). The quad-tree structure is based on a nesting of grid elements of varying sizes, which tends to reduce data redundancy and storage requirements over a fixed-grid, raster structure. Even with a quad-tree based GIS, data must be stored at the finest resolution that is necessary to study detailed processes (Burrough, 1986). However, a quad-tree-type structure may be an effective hierarchial scheme for aggregation/disaggregation of multi-scale data.

Locational analysis

Once the coarsest resolution level of a hierarchical GIS database has been established, it can be used for locational analysis, specifically for locating field plots for sampling ecological variables. Elevation, soil and vegation type data layers are often useful for locating study plots in a manner that representatively stratifies the ecological diversity of the region. Coarse resolution elevation data covering large areas may be readily obtained: (1) from existing digital elevation models (DEM), (2) by digitizing small-scale topographic maps, or (3) by generating a DEM from small-scale stereoscopic aerial photographs or stereoscopic SPOT high resolution (HRV) panchromatic images. Similarly, small-scale GIS coverages of soil distributions can be generated by digitizing soil association maps and vegetation distributions through computer-assisted classification of satellite multi-spectral image data (Stow *et al.*, 1989).

Locational analysis using GIS has to date been used more often for resource management rather than ecological research purposes (Kessel and Cattelino, 1978; Yool *et al.*, 1985; Wells and McKinsey, 1990). Kessel and Cattelino (1978) incorporated GIS and satellite image data with fire-spread models to locate lands of particularly high fire risk. In another example of GIS for fire management, Wells and McKinsey (1990) found that maps of prescribed burn ratings (i.e. assessment of ecological need versus recreational fire danger) for a government managed park were more efficiently produced by using a GIS than by manual methods. The GIS approach was also shown to be more effective, as many more data inputs could be incorporated into the burn-rating process. Burn managers have adopted the technology and are now locating and scheduling prescribed burns within the park with the aid of GIS.

Support remote sensing analysis

Certain data layers in a GIS such as terrain, soil and vegetation type can be useful for improving information extraction capabilities from remotely sensed data (Strahler, 1980; Hallada *et al.*, 1981). For example, the effects of slope and aspect on satellite-derived spectral vegetation indices can be minimized by incorporating digital terrain data with image-processing algorithms that attempt to normalize terrain-related variations (Justice *et al.*, 1980).

Researchers have used GIS to aid in satellite-based mapping of land cover for semi-arid landscapes. Lacaze and Debussche (1984) adopted a GIS approach for selecting training sites based on generalized maps of vegetation physiognomic classes. In spite of the improvements yielded from this approach, they found it difficult to discriminate spectral signatures of broad Mediterranean vegetation cover classes extracted from Landsat/MSS (80 m) data. Graetz *et al.* (1986) developed an image (i.e. raster) based GIS for the semi-arid grasslands of southern Australia. Incorporation of range-type and terrain data in Landsat/MSS identification of land cover classes yielded favourable results. The land cover map then became an integral layer in the GIS.

Geocoded and terrain corrected remotely sensed image data are useful as a graphical backdrop when displaying and analysing computerized GIS coverages. An example of this for a portion of the Jornada Desert in New Mexico, USA, is shown in Figure 2.1. A soils coverage from a vector-coded GIS was displayed over SPOT-HRV multi-spectral image data. The combined soils and image data assisted in locating study plots for relating the proportion of desert shrub cover estimated from aerial photographs to SPOT-derived spectral vegetation indices.

Geographic information systems can also be used to assist in verifying the results of remote sensing and image processing. Figure 2.2 illustrates a methodology for estimating regional net primary production (NPP) from satellite-derived spectral vegetation indices. Terrain data from the GIS are incorporated in a kriging routine

Figure 2.1. SPOT-HRV Band XS2 subscene of Jornada Desert, New Mexico, USA, with soil association boundaries overlaid as a vector GIS coverage.

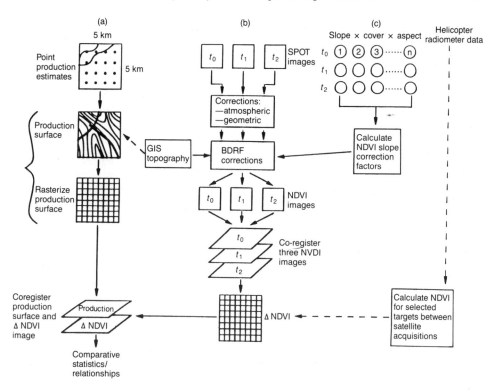

Figure 2.2. Scheme to show the procedures for generating maps of net primary production (NPP) from satellite spectral vegetation index data, and for testing their relative accuracy. Satellite-derived maps are compared with maps generated by GIS-guided interpolation of plot measurements of NPP.

for interpolating a continuous NPP surface from ground measurements of NPP within sample plots. The interpolated surface can then be compared with the satellite-derived map of NPP to assess relative accuracy over large regions. Wickland (1989) provides an overview of why regional estimates of terrestrial ecological variables such as NPP are important, and what role remote sensing will play in making such estimates in the future.

While the present chapter emphasizes the role of GIS in supporting remote-sensing analyses, it should be pointed out that the roles can also be reversed. The research by Graetz *et al.* (1986) described above illustrates the two-way flow of information between remote sensing and GIS. Geocoded and terrain corrected images are useful for deriving GIS coverages of some environmental variables, and as a means for updating dynamic layers of a GIS (Gernazian and Sperry, 1989). Stow *et al.* (1990) found that merged SPOT panchromatic–Landsat thematic mapper image data provided the highest accuracies for updating an outdated vector-coded land-use coverage of an area, which consisted of urban land uses expanding into rural lands predominantly covered with chaparral vegetation. The methods for updating vector-coded coverages used by Stow *et al.* (1990) and the examples shown in Figure 2.1 were made possible by recent commercial advances in the graphical integration of raster and vector data (Gernazian and Sperry, 1989).

Spatial statistical analysis

The spatial encoding structure and large number of data elements inherent in GIS makes them particularly amenable to supporting spatial statistical analyses. Most ecological applications of such analyses have been to assess the spatial interrelationship between environmental variables. Lepart and Debussche (1980) found information efficiency measures useful for establishing relationships between vegetation and other ecological variables in woodland oak communities in southern France. A GIS approach to ecological land classification was used by Davis and Dozier (1988). They used mutual information analysis (specifically, spatial entropy techniques) in order to classify accurately ecological units in Mediterranean landscapes in southern California. Evans *et al.* (1990) found a high degree of spatial correlation between terrain, snow depth and arctic tundra vegetation types using a hierarchical GIS for a portion of the North Slope of Alaska.

    While it may be useful to assess spatial correlations between spatially distributed variables for purposes of exploratory research, a GIS can be exploited for spatial statistical analyses which compare ecological structure between positions in a landscape or between regions. Spatial frequency analyses such as spectral, fractal or hierarchical variance analyses can be applied to one- or two-dimensional data series extracted from GIS layers (Townshend and Justice, 1988; Weiler and Stow, 1990). Lacaze *et al.* (1983) used vegetation maps derived from aerial photographs and Landsat/MSS data to assess surface cover variability on a variety of spatial scales and for a number of seasonal dates. The Mediterranean woodlands that they studied were found to exhibit great heterogeneity at most spatial scales.

Input/output for ecosystem models

Geographic information systems hold much promise for supporting numerical modelling of spatially distributed ecosystem processes. GIS databases can efficiently supply input state variable and model parameter data. Their data structures can also provide an effective structure for partitioning space for model calculations and for storing results.

    There are a number of ways that GISs and ecosystem models can be integrated for ecological studies. Three integration approaches are illustrated in Figure 2.3 using a raster structure (Reynolds and Tenhunen, 1988), and can be generally categorized as:

1. using the GIS to summarize representative or 'average' conditions which are used as parameters or states for patch (i.e. metre square) models;
2. using the GIS to specify parameters and states for model simulations of each patch, with no interaction between patches; and
3. using the GIS to specify parameters and states for model simulations of each patch, with interaction between patches.

    An example of how a GIS can be interfaced with a spatially distributed model of landscape processes is shown in Figure 2.4. Soil erosion rates were estimated using the wind-erosion equation developed by Woodruff and Skiddoway (1965). Erosion estimates were based on soil type and terrain characteristics (slope and aspect) extracted from a vector-coded GIS and data on prevailing wind fields, for a

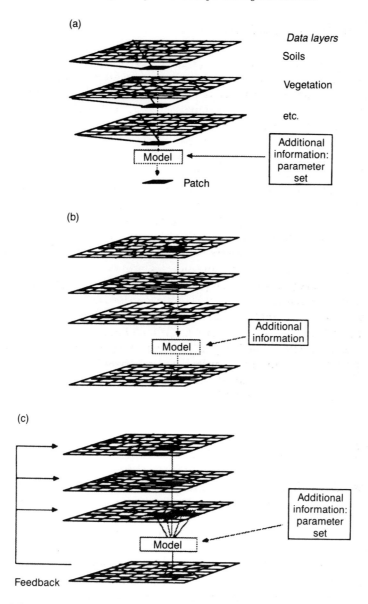

Figure 2.3. Scheme to show three ways in which raster GIS and ecosystem models can be integrated. (a) Patch (or point) model—model runs on average information. (b) Model runs on information of high spatial resolution. Each patch is uniquely defined by GIS data layers. No interactions between grid elements are considered. (c) As (b), but interactions between grid elements are considered.

portion of the Jornada Desert. While the model is static (i.e. no attempt is made to model the transport of wind-eroded material over a landscape), it is capable of exploiting information on the distribution of soil and terrain over a landscape. This type of static GIS overlay modelling has been the one most commonly used to date in GIS modelling of landscape processes.

Haber and Schaller (1988) have integrated spatially distributed ecosystem models with a vector-coded GIS to simulate the influence of land-use practices on

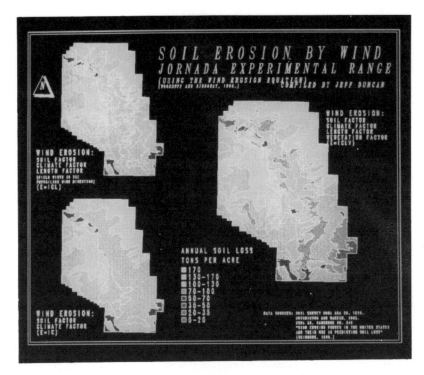

Figure 2.4. Vector GIS coverages of wind-erosion rates for a portion of the Jornada Desert, estimated using the US Department of Agriculture wind erosion equation and GIS overlay of environmental data. The three maps are of coverages generated using different combinations of environmental data.

homogeneous ecosystem units called 'ecotopes', and over heterogeneous mixtures of ecotopes called 'ecochores'. An ecological balance model (EBM) was developed to simulate the transport of materials (e.g. water and nutrients) and energy over an alpine forest landscape. The spatially explicit models were based directly on vector and triangular irregular network (TIN) data structures (Peuker and Chrisman, 1975).

A 'dynamic GIS' provides the basis for simulation modelling of dynamic (i.e. time-dependent) landscape processes (Reynolds and Tenhunen, 1988). Multiple GIS coverage can be used to represent spatially distributed conditions at multiple points in time. Researchers in the hydrological sciences are directly incorporating GIS data structure for dynamic watershed modelling of run-off and soil moisture status. Silfer *et al.* (1986) have directly integrated land cover and terrain data from a vector-coded GIS with a TIN structured hydrological model of overland flow and interflow. Water was routed from each TIN facet to the one or two adjoining downslope facets. Disturbance scenarios can be tested with such a model structure by modifying the land-cover or terrain properties of specific facets.

Conclusions

Landscape ecology and GISs are two fields that are understandably receiving a great deal of attention from researchers and practitioners of a number of disciplines.

The researchers and practitioners affiliated with landscape ecology have for a few years now realized the potential of incorporating GIS into their studies, but are just beginning to exploit this powerful spatial data-handling technology.

This chapter has explored the roles that GIS has played, or soon will play, in advancing knowledge of ecological functioning and human impacts on ecological structure, both on the landscape and on a larger scale. These roles are to:

1. provide a data structure for efficiently storing and managing ecosystems data for large areas;
2. enable aggregation and disaggregation of data between multiple scales;
3. locate study plots and/or environmentally sensitive areas;
4. support spatial statistical analysis of ecological distributions;
5. improve remote-sensing information-extraction capabilities; and
6. provide input data/parameters for ecosystem modelling.

The realization of each of these potential roles presents a major research challenge for the scientists and technicians of the 1990s.

Acknowledgements

Messrs Jeff Duncan, Timo Luostarinen and David McKinsey provided the GIS and remote-sensing examples used in Figures 2.1 and 2.4. Allen Hope and Walt Oechel assisted in the creation of Figure 2.2. James Reynolds and Allen Hope assisted in the creation of Figure 2.3.

References

Allen, T. F. H. and Hoekstra, T. W., 1984, Interlevel relations in ecological research and management: some working principles from hierarchy theory, *Forest Service General Technical Report RM-110*, Fort Collins, CO: US Department of Agriculture.

Allen, T. F. H. and Starr, T. B., 1982, *Hierarchy: Perspectives for Ecological Complexity*, Chicago, IL: University of Chicago Press.

Burrough, P. A., 1986, *Principles of Geographical Information Systems for Land Resources Assessment*, Oxford: Clarendon Press.

Chen, Z. T. and Peuquet, D., 1985, Quad-tree and spectra guide: a fast spatial heuristic search in a large GIS, in *Proceedings Auto-Carto 7*, Washington, DC, pp. 75–83.

Delcourt, H. R., Delcourt, P. A. and Webb III, T., 1983, Dynamic plant ecology; the spectrum of vegetation in space and time, *Quaternary Science Review*, 1, 153–175.

Davis, F. W. and Dozier, J., 1988, Information analysis of a spatial database for ecological land classification, *Photogrammetric Engineering and Remote Sensing*, 56(5), 605–613.

Evans, B. M., Walker, D. A., Benson, C. S., Norstrand, E. A. and Petersen, G. W., 1990, Spatial interrelationships between terrain, snow distribution and vegetation patterns at an Arctic foothills site in Alaska, *Holarctic Ecology*, 12, 2227–2236.

Gernazian, A. and Sperry, S., 1989, The integration of remote sensing and GIS, *Advanced Imaging*, 4(3), 30–33.

Graetz, R. D., Pech, R. P., Gentle, M. R. and O'Callaghan, J. F., 1986, The application of Landsat image data to rangeland assessment and monitoring: the development and demonstration of a land image-based resource information system (LIBRIS), *Journal of Arid Environments*, **10**, 53–80.

Haber, W. and Schaller, J., 1988, Spatial relations among landscape elements quantified by ecological balance methods, *Proc. VIII International Symposium on Problems of Landscape Ecological Research*, Zemplinska Sirava, CSSR, October 1988.

Hallada, W. A., Mertz, F. C., Tinney, L. R., Cosentino, M. J. and Estes, J. E., 1981, Flexible processing of remote sensing data through integration of image processing and geobased information systems, *Proc. 15th International Symposium on Remote Sensing of Environment*, pp. 1099–1112.

Jensen, J. R., 1986, *Introductory Digital Image Processing—A Remote Sensing Perspective*, Englewood Cliffs, NJ: Prentice-Hall.

Justice, C. O., Wharton, S. W. and Holben, B. N., 1980, Application of digital terrain data to quantify and reduce the topographic effect on Landsat data, *NASA Report TM81988*, Greenbelt, MA: NASA.

Kessel, S. R. and Cattelino, P.J., 1978, Evaluation of a fire behavior information integration system for southern California chaparral wildlands, *Environmental Management*, **2**, 135–159.

Lacaze, B. and Debussche, G., 1984, Integration of multiple thematic data with Landsat: some results about the feasibility of Mediterranean land cover inventories, *Proc. Integrated Approaches in Remote Sensing Symposium*, Guildford, UK, pp. 31–39.

Lacaze, B., Debussche, G. and Jardel, J., 1983, Spatial variability of Mediterranean woodlands as deduced from Landsat and ground measurements, *Proc. International Geoscience and Remote Sensing Symposium*, **II**(4), 1–5.

Langley, P. G., 1969, New multi-stage sampling techniques using space and aircraft imagery for forest inventory, *Proc. Sixth International Symposium on Remote Sensing of Environment*, **II**, 1179–1192.

Lepart, J. and Debussche, M., 1980, Information efficiency and regional constellation of environmental variables, *Vegetatio*, **42**, 85–91.

Maffini, G., 1987, Raster versus vector data encoding and handling: a commentary, *Photogrammetric Engineering and Remote Sensing*, **53**, 1397–1398.

Marble, D. F., Calkins, H. W. and Peuquet, D. J., 1984, *Basic Readings in Geographic Information Systems*, Williamsville, NY: SPAD Systems Ltd.

Peuker, T. K. and Chrisman, N., 1975, Cartographic data structures, *American Cartographer*, **2**(1), 55–69.

Reynolds, J. F. and Tenhunen, J. D., 1988, *Response, Resistance and Resilience to, and Recovery from, Disturbance in Arctic Ecosystems. Phase II Conceptual Framework*, Washington, DC: US Department of Energy.

Risser, P. G., 1986, *Spatial and Temporal Variability of Biospheric and Geospheric Processes: Research Needed to Determine Interactions with Global Environmental Change*, Paris: ICSU Press.

Risser, P. G. and Treworgy, C. G., 1985, Overview of Ecological Research Data Management, in *Research Data Management in the Ecological Sciences*, Columbia, SC: University of South Carolina Press.

Silfer, A. T., Kinn, G. J. and Hassett, J. M., 1986, A geographic information system utilizing the triangulated irregular network as a basis for hydrologic modeling, *Proc. Auto-Carto 8*, Baltimore, MA, pp. 129–136.

Steele, J. H., 1985, A comparison of terrestrial and marine ecological systems, *Nature*, **313**, 355–358.

Stow, D., Burns, B. and Hope, A., 1989, Mapping arctic tundra vegetation types using digital SPOT/HRV-XS data: a preliminary assessment, *International Journal of Remote Sensing*, **10**(8), 1451–1457.

Stow, D., Westmoreland, S., McKinsey, D., Mertz, F., Collins, D., Sperry, S. and Nagel, D., 1990, Raster–vector integration for updating land use data, *Proc. 23rd International Symposium on Remote Sensing of Environment*, Bangkok, Thailand, May 1990, pp. 837–844.

Strahler, A. H., 1980, The use of prior probabilities in maximum likelihood classification of remotely-sensed data, *Remote Sensing of Environment*, **10**, 135–163.

Swanson, F. J. and Franklin, J. F., 1988, The Long-Term Ecological Research Program, *EOS*, 32–46.

Townshend, J. R. G. and Justice, C. O., 1988, Selecting the spatial resolution of satellite sensors required for global monitoring of land transformations, *International Journal of Remote Sensing*, **9**(2), 187–236.

Walker, D. A., Binnian, E., Evans, B. M., Lederer, N. D., Nordstrand, E. and Webber, P. J., 1990, Terrain, vegetation and landscape evolution of the R4D research site, Brooks Range Foothill, Alaska, *Holarctic Ecology*, **12**, 238–261.

Weiler, R. A. and Stow, D. A., 1990, Characterizing spatial scales of remotely sensed surface cover variability, *International Journal of Remote Sensing*, **12**(11), 2237–2261.

Wells, M. L. and McKinsey, D. E., 1990, Using a geographic information system for prescribed Fire Management at Cuyamaca Rancho State Park, California, in *Proc. GIS '90 Symposium*, pp. 87–93.

Wickland, D. E., 1989, Future directions for remote sensing in terrestrial ecological research, in Asrar, G. (Ed.), *Theory and Application of Optical Remote Sensing*, New York: Wiley.

Woodruff, N. P. and Skiddoway, F. H., 1965, A wind erosion equation, *Proceedings of the Soil Science Symposium*, 603–609.

Yool, S. R., Eckhardt, D. W., Estes, J. E. and Cosentino, M. J., 1985, Describing the brushfire hazard in southern California, *Annals of the Association of American Geographers*, **75**, 417–430.

3

Landscape ecology, geographic information systems and nature conservation

P. B. Bridgewater

Introduction

Risser *et al.* (1984) state, as one of the conclusions to a workshop on landscape ecology, that

> ... clearly enunciating principles of landscape ecology will catalyse a convergence of existing methodology and theory, and will provide practical improvements in existing methodologies, such as inserting ecological processes more forcefully in *geographic information systems* used for planning purposes. (emphasis mine)

The aim of the present chapter is to discuss how these linkages can be made, particularly in an Australian context.

The same authors give a definition of landscape ecology:

> Landscape ecology considers the development and dynamics of spatial heterogeneity, spatial and temporal interactions and exchanges across heterogeneous landscapes, influences of spatial heterogeneity on biotic and abiotic processes and management of spatial heterogeneity.

This is a very good summary of the differences between landscape ecology and the more species focused approaches of other areas of ecological study. Although this definition does not specifically mention soils or landforms, implicit in it is that interactions between the biota and landforms are the fundamental basics to which landscape ecology should address itself.

Nature conservation involves two distinct but related objectives. The first of these objectives is the maintenance of the maximum degree of biodiversity. The second objective is the development, management and maintenance of ecological infrastructure through the management of protected areas.

Biodiversity maintenance is the 'typical' role of conservation, and is increasingly seen as the prime conservation function. Habitat loss and degradation are the most important causes of biodiversity reduction, but pollution, introduction of exotic species (and, in some systems, overharvesting) all contribute to the reduction. Global warming may well exacerbate the loss and degradation of biodiversity through a variety of effects. Such effects could include the movement of latitudinal gradients so far up mountains that certain communities simply have no niche space left to occupy. Other more obvious effects are the inundation of coastal communities due to sea-level rise.

It is important to note that 'biodiversity' includes conservation of all species, the genetic variability which they contain and the ecological communities they form (McNeely *et al.*, 1990). It is therefore a more encompassing term than simply 'genetic diversity', which has been the buzz-word for the last decade or so. But species and genetic diversity are only one side of the conservation coin. If the aims of species diversity conservation are followed logically, it is possible to see that nature conservation, in those terms, could be most effectively catered for by an increased network of botanical and zoological gardens!

Such a view is, of course, largely nonsense—but it serves to emphasize that biodiversity on its own is not an adequate statement of conservation objectives. It is ecological infrastructure which allows biodiversity to occur, maintain and *change* within the wider environment. Landscape ecology can help maintain biodiversity through an understanding of the structure and function of landscapes.

Good nature-conservation management requires a basic understanding of ecological science at all levels, especially focusing on the landscape ecological aspects. Species and community ecology can easily be dealt with at a local level. If, however, an attempt is made to develop an understanding of ecological infrastructure, the ability to document information and develop models over large areas is vital. This is where geographical information systems (GIS) and their power and potential come into play.

Nature conservation is best served by GIS when there are adequate databases of biological, geological and pedological information, linked through a sound cadastral base. Nature conservation has always made great use of maps, e.g. maps showing species distribution, maps showing the distribution of reserves, and maps showing the distribution of communities (particularly vegetation). But the old fashioned techniques of cartography (involving the very considerable expense of production) can now be replaced by GIS techniques.

In one sense the value of GIS is in the production of 'disposable' maps, produced for an instant, used to solve management issues and then, perhaps, disposed of. The very use of the map will in itself generate a new and different map and thus the interactivity of GIS comes into full play. This chapter attempts to explore how landscape ecology and an understanding of its concepts can interact with the potential of GIS to produce better systems for nature-conservation management.

Historical development of landscape ecology

Landscape ecology has a long history in Europe (see Naveh and Lieberman, 1983) and a very recent history in North America (see Forman and Godron, 1986). Varying approaches have been taken, particularly in Europe, North America and other anglophone countries, notably Australia. Zonneveld (1979) has given a good basic description of landscape ecology and has used different terms for the basic descriptive elements of landscapes.

> *Ecotope.* The smallest holistic land unit, characterised by homogeneity of at least one land attribute of the geosphere (atmosphere, vegetation, soil, rock, water, etc.) and with 'non-excessive variation' in other attributes.
>
> *Land facet.* A combination of ecotypes forming a pattern of spatial relationships, being strongly related to properties of at least one attribute, normally land-form.

Land system. A combination of land facets to form one convenient mapping unit on a reconnaissance scale.

Main landscape. A combination of land systems in one geographical region (also termed the 'macrochore').

In this approach Zonneveld drew on the European system. However, he also took particular note of the systems which had been developed in Australia since the 1950s, mainly by the Commonwealth Scientific and Industrial Research Organisation (CSIRO). In particular, Christian and Stewart (1953) developed a whole series of landscape classifications by which soils, vegetation and land-form were combined into readily observable and easily definable land units. These units were then used in the basic early mapping approaches of the Australian landscape. This approach uses a variety of systems, the base of which is 'the site'—a part of the land surface which, for all practical purposes, is uniform in its land-form, soil and vegetation. In this sense the 'site' is synonymous with the 'ecotope'.

Sites are combined at the second hierarchical level as the 'land unit', which is a group of related sites with a particular land-form. Wherever this land unit occurs it has the same combination of sites. The delineation of a land unit in this system is mainly determined by land-form. Land units of a similar kind are aggregated into a 'land system' which is made up of geomorphologically and geographically associated land units which form recurrent patterns. The boundary of these recurrent land systems usually coincides with some discernible geological or geomorphogenetic feature. Land systems that are the same contain the same land units.

This system proved quite useful for the very broad surveys which were being carried out in Australia over much of the semi-arid regions in the 1950s and 1960s (e.g. Perry, 1960; Stewart *et al.*, 1970). We are now much more aware of the detailed variation which occurs in plant and animal distribution, as well as the distribution of various soil types and the dynamic nature of landscapes. There is thus the potential to build on this valuable early work and produce enhanced descriptive frameworks of landscape structure and function that are of direct relevance to nature conservation.

The ecotope, cited by Zonneveld (1979) as the basic unit (see Forman and Godron, 1986), is usually synonymous with the 'biotope' (Neef, 1967; Agger and Brandt, 1984). Naveh and Lieberman (1983) have also used 'ecotope' as the basic unit of landscape ecological study—being the smallest unit of concrete bio- and techno-ecosystems. In their terms, bio-ecosystems are those maintained by solar energy and natural biotic and abiotic resources, while techno-ecosystems depend on the technological conversion of human, animal and, especially, fossil energy. By emphasising the linkage of systems with their ecotype definitions, these authors introduced a more dynamic element to the statistically identified elements, used by planners, ecologists, etc. It is this move towards dynamic aspects for which modern landscape ecology must strive.

Forman and Godron (1986) have opted for the term 'landscape element', which is a practical highly flexible unit of landscape description—although they have further subdivided to the term 'tessera' to describe the most homogeneous portions of their landscape elements. In many ways their term 'landscape element' is the neatest, least ambiguous term used in recent literature. 'Ecotope', although well established, has the disadvantage of being used in subtly different ways by different schools of landscape ecology. The subtle differences are not helped by differences in the native tongues used by proponents of the different schools.

It is advocated here that the term 'landscape element' should replace 'ecotope'. Landscape elements could be more precisely defined as the basic unit of landscape which is more or less homogeneous, but which is part of a 'dynamic system'. Landscape elements are, of necessity, snapshots in time, and as the burgeoning literature on ecosystem dynamics tell us, their destiny is not necessarily obvious.

Westhoff (1971), who looked at landscapes in terms of their degree of 'naturalness', suggested the following categories of landscape type.

Natural. Landscapes unaffected by human actions, with flora and fauna spontaneous.

Subnatural. Landscapes which, if human activity were removed, would revert to a natural state, with largely spontaneous flora and fauna.

Semi-natural. Landscapes drastically modified by human activity, with the vegetation formation different from potential natural vegetation, but with a considerable degree of natural elements left intact.

Agricultural. Landscapes predominantly arranged by human activity that have no areas of naturalness left or with a great many ruderal and neophytic[1] species.

Van der Maarel (1975) extended this further by incorporating the data of Sukopp (1972), who measured the direction of change from natural to agricultural on the basis of native species loss and neophyte gain.

Van der Maarel (1975) used the terms 'natural', 'near-natural', 'semi-natural', 'agricultural', 'near-cultural' and 'cultural'. The main direction of change is the increasing agriculturisation of landscapes with consequent direct human creation of niches for ruderal and ephemeral species. Occasionally, aggressive neophytes are able to change drastically parts of semi-natural elements—e.g. *Ulex europaeus* in Australia and New Zealand, and *Mimosa pigra* in northern Australia. The former species is usually confined to areas disturbed by human activity and/or physical disturbance, such as stream sides, railway embankments, etc., while the latter is also related to human induced changes, but particularly those accompanying Buffalo introduction.

In a recent study (Bridgewater, 1988) I used the term 'synthetic' to describe vegetation which has gained significant neophytes, but lost relatively few native species (although the proportions of native species in the community may change). Such vegetation is metastable, and is most obvious in the newer settled regions of countries such as Australia, South America, South Africa and western North America, although it can be found on all continents.

Marrying these various terminologies, and recognising the need for unambiguous terms, I suggest that the basic unit for landscape ecology should be the 'landscape element', which can be qualified as 'natural', 'subnatural', 'semi-natural', 'agricultural' or 'synthetic'. Combinations of these elements gives rise to descriptions of landscapes of varying conservation value and potential.

Landscapes in space

One of the central research themes now emerging in landscape ecology concerns the spatial relationship between landscape elements. Major areas of interest include connectivity between landscape elements and associated phenomena such as percolation. Forman and Godron (1986) and Schreiber (1988) give good discussions of connectivity and its emerging importance.

'Connectivity' is a term used to describe an attribute of landscape structure and function. Landscapes are mosaics of patches (elements) linked by corridors or edge fusion. Connectivity is one characteristic of corridors or networks. There are mathematical models of connectivity based in transport geography, but there are also qualitative aspects to connectivity which help in the understanding of landscape structure and function.

Much research effort into connectivity has occurred in the northern hemisphere, with considerable emphasis on the role of anthropic structures such as hedgerows or fence rows. Connectivity is also seen as a tool for understanding the dynamics of metapopulations (e.g. Merriam, 1984). Tropical and southern-hemisphere countries often do not have such anthropic structures as a major constituent of their landscapes. Nonetheless, all landscapes, wherever they sit on the scale of naturalness, have connectivity as a property of their inherent networks.

Connectivity is a matter of scale: connections between landscape patches exist at both the macro and the micro scale. At the local level the term 'ecoline' is proposed to cover any corridor, natural or otherwise (Bridgewater, 1987). At a continental level, corridors between macrochores are recognisable, in a broad sense, and for these the term 'geoline' is proposed.

Geolines are simply intended to indicate trajectories along which biota, energy and nutrients are able to flow around or between major landscape zones. Such trajectories are found *through homogeneous* macrochores, or *between* macrochores (Zonneveld, 1979).

Apart from the land-based views of landscape analysis, mobile organisms have a three-dimensional spatial distribution, as well as dynamic interactions with changing habitats. Boyd and Pirot (1989) tackle one aspect of this in their useful collection of reviews on waterbird flyways. Such flyways are demonstrations of the three-dimensional nature of landscapes and the dynamism of individual species.

At the local landscape level, ecolines are prominent features. Fire-related landscape elements (regeneration patches) are major determinants for ecolines in Australian landscapes. In well-watered regions, stream lines are prominent ecolines, although the high levels of moisture present in these environments may make them unsuitable as routeways for many species. In semi-arid regions, dry watercourses usually tend towards high indices for circuitry—as well as providing numerous isolated 'islands' of dry-land vegetation between their anastamosing branches. Bridgewater (1987) provides some specific examples derived from colour aerial photography.

Connectivity tends to be lower in natural landscapes than in semi- or subnatural landscapes. Ecolines can be detected in all three landscape types. A knowledge of ecolines and their linkages forms a valuable part of developing management strategies for conservation and sustainable land use.

Landscapes in time

Australian landscapes, while appearing superficially highly stable and unchanging, are an example where a knowledge of dynamics is vital in understanding their potential. Climate is a major controlling factor, with natural effects such as droughts

(and associated fire), cyclones and seasonal rainfall shifts all being important ecological controllers.

   In the central part of Australia, cyclones which cross the eastern, northern and
western coasts act as occasional but major influences on the biota. Because cyclones
are unpredictable in their frequency and their likely trajectory, no estimate can be
made of the climate of much of central Australia on an annual basis. Purdie (1984)
notes:

> The cycles of fire and drought/wet have probably operated for thousands of
> years, and still maintain the vegetation in a dynamic state resulting in spatial
> and temporal heterogeneity of community structure and composition which are
> not apparent at a single place or time.

Such perturbations in an otherwise steady-state system are reflected in an
occasional dramatic change in landscape elements. This change is effected by the
filling of ephemeral watercourses and lakes and mirrored by ephemeral populations
of plants and animals which persist only when conditions are favourable. Season-
ality and the amount of rainfall will determine which organisms develop and their
level of persistence (e.g. Mott, 1973). Whichever ephemeral plant community is produced, it is within the persistent structural framework provided by the perennial
communities.

   Main (1976) has identified the importance of fire: 'as an ecological factor in arid
Australia, fire is as ubiquitous as drought'. The role of fire in shaping Australian
landscapes has been receiving increasing attention, particularly in view of the potential use of fire as a management tool.

   Fire is a major factor in creating and maintaining landscape elements. Fire has
also been responsible for creating major shifts in vegetation patterns (see Noble and
Slatyer, 1981). Particular examples include the *Gymnoschoenus* sedgeland, *Nothofagus* temperate rainforest and the sclerophyllous forest landscape in western Tasmania. Jackson (1968) has described these changes in regenerating vegetation in
similar terms to the genetic-drift concept used by population geneticists. The term
'ecological drift' coined by Jackson neatly describes the flux between these three
vegetation types. Fire frequency determines the waxing and waning of the three
types, with low fire frequencies favouring *Nothofagus*, while high frequencies (12–25
years) favour the *Gymnoschoenus* sedgeland.

   Even intense wildfires produce a patchy landscape—small relatively unaffected
islands acting as seed areas for regeneration (Christensen and Kimber, 1975). Fires
in arid areas may have a variety of effects, depending on the time of year of burning.
Summer fires (a natural occurrence in *Triodia* grasslands) control shrub invasions
(Suijdendorp, 1981). Survival and regeneration of the grasses depend on the amount
and timing of the next rains. Winter firing in *Triodia pungens* and *Acacia translucens*
along the Pilbara coast of north-western Australia effectively promotes the increasing dominance of *Acacia*, particularly if aided by sheep grazing following the fire.

   A major complicating factor affecting fire in Australian environments is human
activity. Australian Aborigines used fire as a tool for environmental manipulation.
Major changes brought on by activities following European colonisation in 1788
have been: species extinction, changes in ecosystem structure and function, wider
distribution of exotic plants and animals, and creation of new synthetic communities
(Bridgewater, 1990). Fauna and flora suffered severe disruption in these areas with
the discovery and settlement by Europeans from the late eighteenth century to the
present. Although many such landscapes may appear valueless from the conserva-

tion viewpoint, they may well have reservoirs of biodiversity, and offer connectivity between adjacent semi- or sub-natural landscapes, and so fulfil an important role in nature conservation.

That is not to say the landscapes of Australia were unaffected by people prior to settlement by Europeans. There is increasing evidence that for a period of 40 000–50 000 years Aborigines used fire as the most important tool in modifying their environment by what is best described as 'natural engineering' (e.g. Nicholson, 1981; Haynes, 1985). Thus the present-day landscapes originate from landscapes shaped by fire, both natural and human modified.

GIS: a major tool for landscape ecology

The response to these dynamic ecological effects at the landscape scale has been an increasing desire to focus attention on GIS as a major management tool. Most of the Australian states are actively developing GIS for land-management purposes, and the Australian Government has a number of initiatives in GIS at the national level.

An environmental related GIS is the Environmental Resource Information Network (ERIN) currently under development by the Australian Parks and Wildlife Service to provide a 1 : 1 million coverage, continent wide, of significant environmental information by 1992.

An established GIS which is highly relevant is the Environmental Resource Mapping System (E-RMS) produced by the New South Wales National Parks and Wildlife Service (Ferrier, 1989). E-RMS is fully generic and, therefore, applicable to any new geographical database without requiring software modification. It runs on IBM-compatible personal computers (PCs) and requires no special 'add-on' hardware, except a standard EGA/VGA graphics adaptor and a small hard disk. Data for an E-RMS database can be input by manual encoding, digitising, or imported from other popular GIS/image-processing systems (e.g. PC ARC/INFO and ERDAS).

E-RMS is entirely menu driven and requires minimal user training. The capabilities of E-RMS extend far beyond the standard data storage and retrieval functions offered by most GIS products. The basic philosophy behind the development of E-RMS was to provide an easy to use system for converting raw map-based environmental data into information of direct value to environmental-resource managers. E-RMS therefore offers a wide range of tools for data analysis, model building and extrapolation.

E-RMS is primarily a grid-cell (or raster) based geographical information system. Spatial entities such as points, lines and polygons are represented within E-RMS using *grid-cell encoding*. Although E-RMS also stores line data as precise vectors (for cartographic enhancement of maps), the system's manipulative and analytical tools operate only on grid-cell data.

Any given E-RMS database uses a set grid-cell size (e.g. 100 m). The grid-cell system is aligned with the Universal Transverse Mercator (UTM) or Australian Map Grid (AMG). A database covers a specified rectangular area of grid cells called the *database rectangle*.

Information on a wide range of attributes can be associated with each grid cell in a database. E-RMS uses a strict terminology when referring to this information. The most basic unit is the *feature*. A feature is a property of each grid cell which is coded as being either present or absent (e.g. 'subtropical rainforest', 'steep slope', or 'logged in 1980').

Related features can be grouped to form a *layer*. For example, several features representing different types of vegetation may be grouped within a layer called 'vegetation'. It should be noted that the features in a layer need not be spatially mutually exclusive. For example, a 'rare birds' layer could have both 'marbled frog-mouth' and 'eastern bristlebird' features coded as present in the same grid cell. 'Dynamic' features such as fire patterns can also be included.

Features and layers may be either stored permanently in an E-RMS database, or be created temporarily by manipulating existing database information. Layers stored permanently in the database are also called *variables*, while their features are also called *categories*. The structure of an E-RMS database can, therefore, represent a highly dynamic system for conservation management.

The Australian National Parks and Wildlife Service has used E-RMS as a PC-based GIS for a number of programs including:

1. pattern-based GIS for Kakadu and Uluru National Park management purposes;
2. assessing the representativeness of ecosystem/landscape (domain) types across northern Australia; and
3. investigating species distributions and their relationship with nationally important ecosystems, e.g. koalas (Phillips, 1990).

Pattern-based GIS for Uluru and Kakadu National Park in the Northern Territory has involved work on vegetation types and fire. As noted earlier, fire is a major factor in Australian landscapes and a knowledge of fire behaviour is vital in developing management plans. Figure 3.1 shows management burns and wildfire patterns plotted on the base map for Uluru National Park. The scale on all maps is set automatically to provide the largest plot available for the size of the plotter. Tick marks around the edge of the Kakadu and Uluru maps give some indication of the size of the plots. Such a map shows the relatively disastrous effects of wildfires. Other layers can be superimposed or combined to form the basis for a series of management-related questions. This GIS is under development at present, but obviously the combination of wildfires with vegetation types can form the basis of a strategy for prescribed management burns.

For Kakadu National Park the timing of fire occurrence in the 'early dry' or 'late dry' season can give rise to serious management problems. Here the E-RMS GIS process has been used to great effect. Figure 3.2 shows the preventative burns made early in the 1989 dry season, with the aim of achieving 'barriers' to wildfires which occur in the late dry season. Clearly the planned preventive burns have been ineffective, which stimulated review of procedures and practices to try and produce a better fire-management system. Here the GIS/landscape information is being used interactively to develop an effective and safe fire system.

Kakadu National Park has had the benefit of input to planning from traditional Aboriginal owners, who provided specific input into the fire regimes from their traditional knowledge. The GIS is thus a powerful tool, integrating traditional knowledge, modern park-management techniques and landscape dynamics.

Figure 3.3 shows the distribution of the eight types of rainforest recognised in Kakadu. While this is useful in itself, the 'layer' can be merged with fire data to show potential threats to this high quality and rare vegetation type.

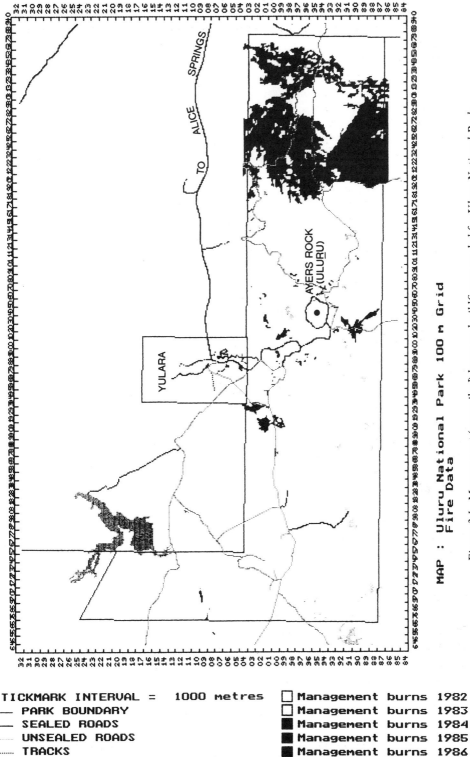

MAP : Uluru National Park 100 m Grid
Fire Data

Figure 3.1. Management (prescribed) burns and wildfires recorded for Uluru National Park.

TICKMARK INTERVAL = 1000 metres

— PARK BOUNDARY
— SEALED ROADS
···· UNSEALED ROADS
······ TRACKS

☐ Management burns 1982
☐ Management burns 1983
■ Management burns 1984
■ Management burns 1985
■ Management burns 1986
▨ Management burns 1987
☐ Wildfires 1983 ?
▨ Wildfires 1985 ?
■ Wildfires 1986 ?

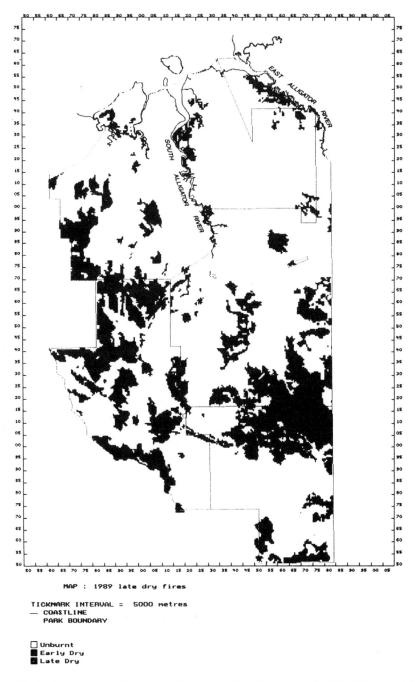

MAP : 1989 late dry fires

TICKMARK INTERVAL = 5000 metres
— COASTLINE
   PARK BOUNDARY

☐ Unburnt
■ Early Dry
■ Late Dry

Figure 3.2. Early dry-season (preventative) burns and late dry-season (wildfire) burns in Kakadu National Park, 1989. Boundaries are internal park-management areas.

Figure 3.4 shows a GIS of environmental domains for northern Australia. Domain analysis is a recently developed technique which is still being refined. Mackey *et al.* (1989) give a description of the procedure and a regional application. Domain analysis attempts to use the primary attributes which control or modify environmental and biological processes—particularly climate and terrain.

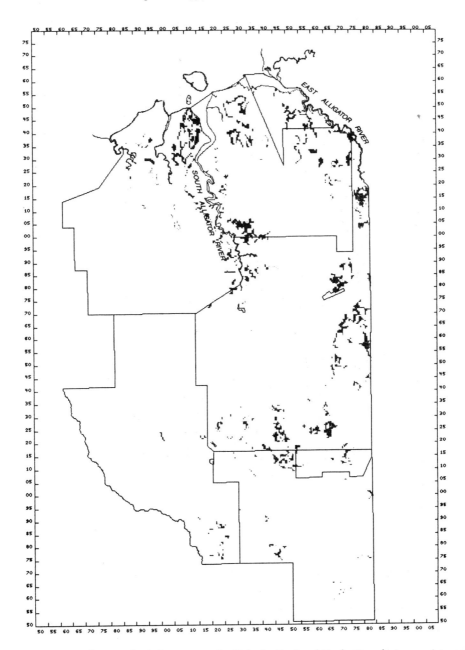

Figure 3.3. Distribution of rainforest types in Kakadu National Park. Boundaries are internal park-management areas.

The methodology exploits the capabilities of GIS technology but is firmly based on concepts of minimum data sets, primary attribute data and close coupling with process-based predictive models. The principal characteristics of the methodology include the following.

1. Emphasis being given to those primary environmental attributes which modulate physical processes and biological responses.

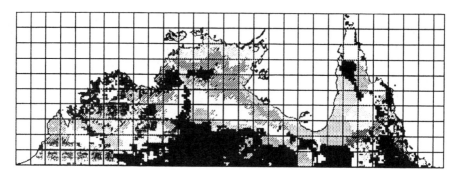

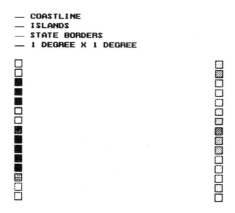

Figure 3.4. Environmental domains for northern Australia.

2. A stable framework for future analysis is provided because it is based on relatively stable attributes of the physical environment.

3. A wide range of potential land uses and environmental-change effects can be evaluated, given the attributes which form the domains. The environmental domains generated are not absolutes. Other classifications may be preferred for a particular purpose and the system makes such alternatives possible.

4. The grids used are finite subdivisions of standard longitude and latitude and can be converted to any scale and projection used in maps so that all outputs can be located accurately in map space.

5. Point data relating to environmental and/or biological data can be allocated to any number of defined zones for subsequent analysis.

These applications are GIS giving 'scenarios' a value for nature conservation. The ability to generate scenarios has always been limited in nature-conservation management: GIS offers great potential. In landscape ecological terms, if the landscape elements can be defined, and the impact can be analysed for each element, then the possibilities for mediating impacts can be explored. The GIS, coupled with a landscape ecological framework, and a knowledge of sensitive keystone species, can therefore offer the potential for posing (and solving) the crucial 'What if?' questions.

The above-described examples all demonstrate the wide range of use that can be expected for GIS in landscape ecology, linked to nature-conservation management. For the future, there is a need to define landscape-ecology concepts as they apply to conservation objectives. Such concepts can then, through the medium of GIS, be made readily available to conservation planners and managers. The ability of GIS to be hosted on PCs and highly 'mobile' computers means that those who need the information on the spot can now be satisfied. As more field-based analysis is done, there will inevitably be feedback to the conceptual development of landscape ecology, and we can look forward to a rapidly evolving science.

Note

[1] Neophytes are species occurring in landscape since 1500 A.D., having been either deliberately or accidentally introduced by human activity.

References

Agger, P. and Brandt, J., 1984, Registration methods for studying the development of small-scale biotope structures in rural Denmark, in Brandt, T. and Agger, P. (Eds), *Proc. First International Seminar on Methodology in Landscape Ecological Research and Planning*, Vol. II, pp. 61–72, Roskilde, Denmark: International Association of Landscape Ecology.

Boyd, H. and Pirot, T. Y. (Eds), 1989, Flyways and reserve networks for water birds, *IWRB Special Publication*, **9**, Slimbridge: IWRB.

Bridgewater, P. B., 1987, Connectivity: an Australian perspective, in Saunders, D. A., Arnold, G. W., Burbidge, A. A. and Hopkins, A. J. M. (Eds), *The Role of Remnant Vegetation in Nature Conservation*, pp. 195–200, Sydney: Surrey Beatty.

Bridgewater, P. B., 1988, Synthetic plant communities; problems in definition and management, *Flora*, **180**, 139–144.

Bridgewater, P. B., 1990, The role of synthetic vegetation in present and future landscapes of Australia, *Proceedings of the Ecological Society of Australia*, **16**, 129–134.

Christian, C. S. and Stewart, G. A., 1953, General report on survey of Katherine–Darwin region, 1946, *Land Research Series*, **1**, Canberra: CSIRO.

Christensen, P. and Kimber, P., 1975, Effect of prescribed burning on the flora and fauna of south-west Australian forests, *Proceedings of the Ecological Society of Australia*, **9**, 85–107.

Ferrier, S., 1989, *Environmental Resource Mapping System (E-RMS), Users Manual for Version 2.1*, Sydney: National Parks and Wildlife Service of New South Wales.

Forman, R. T. T. and Godron, M., 1986, *Landscape Ecology*, New York: Wiley.

Haynes, C. D., 1985, The pattern and ecology of *munwag*: traditional aboriginal fire regimes in north central Arnhemland, *Proceedings of the Ecological Society of Australia*, **13**, 203–214.

Jackson, W. D., 1968, Fire, air, water and earth—an elemental ecology of Tasmania, *Proceedings of the Ecological Society of Australia*, **3**, 9–16.

Mackey, B. G., Nix, H. A., Stein, J. A., Cork, S.E. and Bullen, F. T., 1989, Assessing the representativeness of the wet tropics of Queensland world heritage property, *Biological Conservation*, **50**, 279–303.

Main, A. R., 1976, Adaptions of Australian vertebrates to desert conditions, in

Goodall, D. W. (Ed.), *Evolution of Desert Biota*, pp. 101–131, Austin, TX: Univeristy of Texas Press.

McNeely, J. A., Miller, K. R., Reid, W. V., Mittermeier, R. A. and Werner, T. B., 1990, *Conserving the World's Biological Diversity*, IUCN Gland.

Merriam, G., 1984, Connectivity: a fundamental ecological characteristic of landscape pattern, in Brandt, T. and Agger, P. (Eds), *Proc. First International Seminar on Methodology in Landscape Ecological Research and Planning*, Vol. I, pp. 5–16, Roskilde, Denmark: International Association of Landscape Ecology.

Mott, J. J., 1973, Temporal and spatial distribution of an annual flora in an arid region of Western Australia, *Tropical Grasslands*, **7**, 89–97.

Naveh, Z. and Lieberman, A. S., 1983, *Landscape Ecology: Theory and Applications*, New York: Springer-Verlag.

Neef, E., 1967, Die Theoretischen Grundlagen der Landschaftslehre, in *Geographisch—kartographische Anstalt Gotha*, Leipzig: Hermann Haack.

Nicholson, P. H., 1981, Fire and the Australian Aborigine—an enigma, in Gill, A. M., Groves, R. H. and Noble, I. R. (Eds), *Fire and the Australian Biota*, pp. 55–76, Canberra: Australian Academy of Science.

Noble, I. R. and Slatyer, R. O., 1981, Concepts and models of succession in vascular plants communities subject to recurrent fire, in Gill, A. M., Groves, R.H. and Noble, I. R. (Eds), *Fire and the Australian Biota*, pp. 311–338, Canberra: Australian Academy of Science.

Phillips, B., 1990, *Koalas*, Canberra: Australian Government Publishing Service.

Perry, R. A., 1960, Pasture Lands of the Northern Territory, Australia, *Australian Land Reseach Series* **5**, Canberra: CSIRO.

Purdie, R., 1984, Land systems of the Simpson Desert region, *CSIRO Division of Water and Land Resources Natural Resources Series* **2**, Canberra: CSIRO.

Risser, P. G., Karr, J. R. and Forman, R. T. T., 1984, Landscape ecology: directions and approaches, *Illinois Natural History Survey Special Publications* **2**, Champaign–Urbana, Illinois Natural History Survey.

Schreiber, K. F. (Ed.), 1988, Connectivity in landscape ecology, in *Proc. 2nd International Seminar of the International Association for Landscape Ecology* (*Munstersche Geographische Arbeiten* **29**). Ferdinand Schöningh, Paderborn, Germany.

Stewart, G. A., Perry, R. A., Paterson, S. J., Travers, D. M., Slatyer, R. A., Dunn, P. R., Jones, P. J. and Sleeman, J. R., 1970, Lands of the Ord–Victoria area, W. A. and N. T., *Australian Land Reseach Series*, **28**, Canberra: CSIRO.

Suijdendorp, H., 1981, Responses of the hummock grasslands of Northwestern Australia to fire, in Gill, A. M., Groves, R. H. and Noble, I. R. (Eds), *Fire and the Australian Biota*, pp. 417–426, Canberra: Australian Academy of Science.

Sukopp, H., 1972, Wandel von Flora und Vegetation in Mitteleuropa unter dem Einfluss des Menschen, *Berichte der Landwirtschaft*, **50**, 112–139.

Van der Maarel, E., 1975, Man-made natural ecosystems in environmental management and planning, in van Dobbin, W. H. and Lowe-McConnel, R. H. (Eds), *Unifying Concepts in Ecology*, pp. 263–274, The Hague: Dr W. Junk.

Westhoff, V., 1971, The dynamic structure of plant communities in relation to the objectives of conservation, in Duffey, E. and Watt, A. S. (Eds), *The Scientific Management of Animal and Plant Communities for Conservation*, Oxford: Blackwell.

Zonneveld, I. S., 1979, *Land Evaluation and Landscape Science*, Enchede, The Netherlands: International Institute for Aerial Survey and Earth Sciences.

PART III

Conceptual Issues

4

The tradition of landscape ecology in Czechoslovakia

J. R. Petch and J. Kolejka

Introduction

The capacity of electronic machines to store and transmit data and the ingenuity of programmers and systems designers to translate that power into useful work have transformed our approaches to spatial data. Mapping information of high levels of complexity, requiring multiple analyses and syntheses of data is now a normal part of the computing repertoire.

At first sight, therefore, it may seem strange to consider the tradition of landscape ecology of an eastern European country in a text dealing also with spatial information systems. The topic is nevertheless signficant at a fundamental level—the level of methodology. The term 'methodology' is used here to mean the logical bases for taking and using measurements, for design of maps, for structuring of models and for interpreting the products of GIS.

It can be appreciated that modern mapping and geographic information system (GIS) software will become essential tools for anyone dealing with environmental planning and management. As with all computing, however, there is the question of what the computing algorithm is doing when it performs a particular task. The method of dealing with empirical data is not a question of computing. It is a question of logic using scientific and management principles or of using professional codes of practice. The significance of landscape ecology for information systems is that it provides a rationale for dealing with the environment at a small scale (large area), in a spatial manner and based on scientific principles. This might seem unexceptional, but a further point is that there are no strong, developed traditions of large area landscape study in the UK or the USA, yet many of the problems which face us fall into this category. Our approach to problems such as land-use change, land-pollution control, water pollution, erosion control, recreation planning, urban sprawl and waste management needs to be on a small scale.

Computers provide excellent tools for presenting data about areas of enormous size (assuming we have data), but how can we use the data to deal with problems in large heterogeneous, complex areas? What logic can we build into our methods? It is here that landscape ecology needs to be considered, and this explains why it might be useful to turn our attention to the experience of central Europe. It has something to teach those in the West about how to approach problems of large areas, about how to handle data on landscapes and, in fact, about what landscapes are. Each of these issues has been neglected in UK and US landscape work and it may well be that we have something to learn. Goodchild (1990) has pointed clearly to the need to develop scientific questions and methods in GIS. The landscape ecology described here may provide a means of achieving this goal. It is only one approach and the objective is not to argue for its superiority or universality, but for a more scientific approach to GIS applications on the small scale (Haines-Young and Petch, 1985).

Development of Czechoslovakian landscape studies

Czechoslovakian ideas and concepts about landscape form part of the central European tradition of landscape studies. Modern Czechoslovakian landscape ecology is similar in methods and objectives to current work in the formerly socialist countries of eastern Europe as well as Germany (West), Holland and Belgium. Schools in each country have developed since the mid-nineteenth century but with different influences at different periods, interrupted by wars and, in Eastern Europe, by socialisation. In addition, Czechoslovakian studies fall into two camps: those of the Czech school, represented by the work of the Academy of Science; and those of the Slovak school, represented by the Slovak Academy of Science, which has developed a strong international reputation.

In central Europe, ideas about landscape had a long pedigree (Kolejka, 1988). In the early nineteenth century Alexander von Humbolt defined 'landscape' as the total character of the earth, including in his concept the idea of landscape as the home of man with cultural and economic, as well as physical, dimensions. His ideas are the parents of all current ideas on landscape in Czechoslovakia and elsewhere.

As early as 1855, natural landscape units were delimited for the Kingdom of Bohemia for the purpose of economic planning. Bohemia at that time was a semi-autonomous part of the Austro-Hungarian Empire. The units were to serve as the basis for organizing increased production on agricultural and forest soils. To achieve this it was considered necessary to pay due regard to topographical, geological, edaphic and climatic conditions. This, it must be stressed, was not at the site level, but at the national level and in a coordinated system as a basis for national- and regional-scale planning. Such an approach is an early recognition not only of the significance of soil and other factors for plant production (see Bunting, 1989), but also of the fact that the physical components of areas have manifestations as mappable units which are operationally important. It recognizes also that there is a spatial structure to landscape at a high level.

After the Czechoslovakian republic was formed in 1918, this work of delineating units continued. Initially the emphasis was on mapping what were called 'natural landscapes' (Kolacek, 1924) or natural areas (Dvorsky, 1918; Kral, 1930), but in effect the delineation was based essentially on political or economic areas and, therefore, most effort concentrated on defining criteria for delineating natural properties within these given units.

In the period of the Republic there were early ideas of hierarchies of units expressed by Dedina (1927). These were based on anthropogenic and administrative as well as physical components and, importantly, considered the degree of homogeneity of landscape areas in constructing hierarchies. Although in defining such units there was inevitably some correspondence with natural landscape areas, the role of physical factors was subconscious and not concrete.

Kral's (1930) 'natural landscape' was an early attempt to define natural units. Although his approach was unacceptably deterministic, in terms of how he viewed man's relation with his environment, he expressed the idea of the creation of landscape units by the people who lived in an area and the dynamic nature of the relation between man and nature. Korcak's (1936) concept of land was more elaborate and advanced. It included ideas of a synthesis of geographical information in defining landscape areas and of the dynamics of the relationships between landscape components.

Czech research was suspended in the period leading up to and during World War II, but Slovak work was not affected to the same degree. Hromadka (1943) continued the development of the ideas of landscape synthesis in Slovakia for both natural and cultural landscapes. In the post-War period his studies formed the basis for the Slovak landscape school, now centred on the Institute of Experimental Ecology, Slovak Academy of Sciencies, Bratislava. It is expressed in an advanced form in the LANDEP method (Ruzicka and Miklos, 1986).

In the Czech lands, the post-War period was characterized by a closer integration of national policy and scientific research. Several institutions were directed towards continuing the tradition of landscape ecology in basic and applied research. The Geography Institute of the Czech Academy of Sciences was established in 1953. It has pursued a series of 5-year plans (Pribyl and Stehlik, 1988; Gardavsky, 1988) the objectives of which were mirrored by research in universities and other institutes in Bohemia. In these studies a major conceptual stimulus (Hribova, 1956) led by Low came from the German geo-ecological school of Schulze, Troll and others and the Soviet school of Isacenko, Zabelin and others. These introduced a strong ecological element to Czech thought and laid the foundations for current ideas and methods of landscape ecology which integrate physical and ecological approaches to landscape study.

Research programmes in the Czech Academy of Sciences have evolved from the early programmes which involved defining systematic complex physico-geographic regions for the whole country at a scale of 1 : 500 000. These formed the basis and framework for later larger scale landscape work. They and their derivative works are represented in part in the Population Atlas of Czechoslovakia, a source of reference and of ideas far richer than any typical Anglo-American atlas—for example, it contains maps of natural environment zones defined on the basis of substrata, mesoclimate relief, zonal soil types, zonal vegetation association and ground water characteristics.

Successive periods of research related to landscape ecology (Pribyl and Stehlik, 1988; Kolejka, 1988; Bucek and Mikulik, 1988) involved, under the State Plan of Technical Development, detailed studies of model regions (Bucek and Mikulik, 1977) as part of a strategy of protecting the environment and for creating new environments. Small- and large-scale studies of soil erosion (Pelisek and Sekaninova, 1975; Hradek, 1988) and the design of biogeographical maps are the types of work which have continued the tradition of analytical studies at the Academy. They are intended to form a firm scientific foundation for synthetic studies of complex geographical areas which is continuing in basic and applied studies at the Academy (Kolejka, 1988) and elsewhere (Hynek, 1981).

The question of the reality of landscape units, which is a fundamental assumption of the landscape approach, has been neglected, and only in the last few decades has it been considered seriously, particularly in the work done at the Czechoslovak Academy of Sciences (Kolejka, 1988).

Conceptual bases for landscape studies

The most important element of ecological ideas about landscape is that a landscape is an amalgam of natural and cultural influences and a manifestation of local physical and economic conditions which has evolved over long periods of time. Such ideas are in their implications altogether richer than the concept of landscape used by mainstream UK and US workers in either geography or ecology. There are exceptions, and today Anglo-American ecologists are becoming more aware of landscape ecology in the European tradition (Naveh and Lieberman, 1986; Risser, 1987). Nevertheless, these countries lack the tradition of the European schools expressed in their conceptual basis, research problems, methodology and attempts to solve practical problems at the landscape level. The works of Forman and Godron (1981), Naveh (1982), Risser (1985), Risser *et al.* (1984) acknowledge the advances made in European studies. Both Forman and Godron (1981) and Risser *et al.* (1984) present principles of landscape ecology which are meant to provide conceptual bases for landscape work. Much of their work and the work of those who regard these ideas as important (a view the author shares) (see, for example, the volumed edited by Turner (1987)) concentrates on issues such as island biogeography, disturbance, landscape heterogeneity and processes of succession and other dynamics. Risser (1987), in his state-of-the-art review, concludes that 'the current status of landscape ecology is typified by the idea of a semi-serious ... portfolio' which could consist of 'several case studies and a number of existing proposed analytical and experimental techniques, all thought to be useful but not organised in a strong conceptual or theoretical manner'. East European landscape ecology contrasts with these works in having a more holistic approach to landscape description and a more coherent 'portfolio' of analyses and syntheses of data.

Both the conceptual and analytical basis for landscape ecological studies is that the contemporary cultural landscape is a complex geographical system formed by both natural and socio-economic factors the structural dynamic relationships between which have evolved over long periods. Ecosystem terms and concepts are employed in landscape ecology, but in combination with terms which describe the landscape as a functional economic system. The idea of a natural ecosystem is important, but not in the same way as for classical Anglo-American schools which are based on the idea of the climax. Odum's (1969) paper on ecosystems can be used as a summary of the thinking of that school. It was an attempt to structure ecological theory in order to, amongst other things, understand change and manage ecosystems. It was based, albeit implicitly, on traditional ideas of natural systems and climaxes whose weaknesses continue to be exposed (Murdoch, 1966; Vitousek and Reiners, 1977). At present, theoretical and empirical studies of succession or change can be considered to be coming to terms with the abandonment of the Clementsian ideas which dominated ecology for most of this century. As a consequence, the idea of an ecosystem as an operational unit is now somewhat discredited. The emphasis in Anglo-American ecology has shifted strongly to studies of function in ecosystems, communities and populations (Roughgarden *et al.*, 1989). Consequently there has been a neglect of the holistic approach amidst quite proper charges that it can be unscientific (Murdoch, 1966). More recently, landscape ecologists have concentrated on the more easily defined problems of patch dynamics, heterogeneity and so on (Turner, 1987). Such studies are essential in the development of landscape ecology, but they do not address the issue of how to deal operationally with landscapes which are difficult to define but which have, nevertheless, real ecological problems.

In European landscape ecology, there has been a continuing role for ideas on natural ecosystems, and on holistic approaches, but these ideas are different from those of Anglo-American ecology in certain important respects. There is an emphasis on ecosystem function, but ideas of climax and succession which relate to smaller scales and longer periods are not of central concern. Rather, natural ecosystems are regarded as functioning units the character of which has been determined by natural conditions and by management. In landscape terms the critical assumptions are not about climax or succession but about the spatial structure of the landscape and about the harmony between vegetation and natural conditions.

The spatial (chorological) structure is described and analysed in terms of its stability. Landscape stability is held to be a function of two factors. The first is the spatial relationships between stabilizing and destabilizing elements. Stabilizing elements in a cultural landscape are ecosystems such as forest, botanically rich grassland and hedgerows. They can be naturally regenerating or managed. They are characteristically species rich, although not always, and are physically stable, i.e. do not show erosion or nutrient loss. Destabilizing elements, such as arable land, are physically unstable and species poor. Spatial relationships between stabilizing and destabilizing elements are expressed in the skeleton of ecological stability (Bucek and Lacina, 1981, 1985). This is a map of the distribution of both types of area. Assessment of the connectivity of stabilizing elements and the isolation of destabilizing elements is the basis for assessing stability. Maps (and stability analyses) can be on different scales, producing hierarchies of stabilizing centres and skeletons. Ideas of biocentres, corridors and patches as functioning elements of the landscape are important for landscape stability assessment, mirroring recent ecological work in the west (West *et al.*, 1980; Forman and Godron, 1986; Turner, 1987).

The 'harmony' between existing vegetation and natural conditions is the second element of landscape stability analysis. This harmony is judged for each spatial unit of an ecosystem. The appropriateness of land cover and land-management practices is set against the optimum, possible and inappropriate uses for areas of land. These hypothetical uses are determined from a consideration of the abiotic characteristics of land: geology, soil, topography, hydrology and microclimate. Deciding on these hypothetical land uses relies on two assumptions. First that units of the landscape can be mapped, and secondly that for any unit a particular use can be deemed optimum, possible or inappropriate.

These mappable units of the landscape are termed 'geocomplexes' or 'geobiocoenoses'. The idea of geocomplexes rests on a number of important assumptions. First, that there are clear, repeated relationships between abiotic factors such that particular combinations of climate, lithology and relief give rise to particular types of soil and hydrological and geomorphical conditions. Second, that these conditions can be mapped as homogeneous, discrete units of the landscape. Third, that a particular geocomplex is characterized by a specific assemblage of processes, in particular those describing external controls on fluxes of nutrients and energy through ecosystems. Thus biogeocoenoses are not statements about vegetation, but about conditions for vegetation growth.

Much of European landscape ecology is based on the idea of landscape units. This issue has been the cause of criticism and scepticism amongst ecologists and other landscape scientists. There are questions about the validity of units and about their supposed functionality. Can reliable statements be made, for example, about their optimum or inappropriate use? In other words, areal units may not be real, definable and concrete spatial entities which have any meaning for predicting behaviour of vegetation. Similar, more familiar criticisms of some significance to landscape ecologists are those about the European schools of vegetation analysis of Bravu-Blanquet, which developed around the turn of the century (Colinvaux, 1973; Whittaker, 1975). The criticisms relate to the methods of vegetation analysis and the nature of the defined units. They are important from the point of view of a theoretical ecology which has a reductionist approach (Harper, 1985). However, they are inappropriate for European landscape studies because they fail to appreciate the cultural landscape which is a mosaic of managed blocks of land developed under man's hand for many centuries, in which most vegetation has been planted, cut and grazed and which was *made* uniform in small areas.

The definition of 'landscape units' has a deeper basis than merely the recognition of vegetation blocks. Clearly, vegetation can be changed drastically and the appearance of the landscape can be altered. But there would still remain characteristics of the landscape, independent of a transient vegetation cover, which would be important, if not critical, in determining its character and function. Geology, aspect, slope, topography, whether or not a site is water receiving or shedding, groundwater conditions and so on are essentially unchanging. These factors change their relevance with scale (Schum and Lichty, 1965), but scale notwithstanding they are parts of the characteristic of any site or area and are mappable. In addition, these characteristics can be used to make statements about the performance of vegetation or managed agricultural and forest systems. The concepts of landscape units include also the idea that landscapes have a structure which determines spatial relationships within and between vegetation patches and landscape units. Such ideas form the basis of analytical and synthetic studies of landscape which are at the heart of the Czechoslovak schools. These are examined in more detail here before examples of the work are described.

A biogeocoenosis is a part of the landscape defined in terms of mapped abiotic characteristics such as soil and topography. It expresses not only the mapped variables but also the implicit relationships between other abiotic components. These could be expressed as other operational variables. So, for example, soil units are supposed to represent particular levels of pH, nutrient status and so on, whilst topographical units represent slope, aspect, flow convergence and so on. The existence of specific biotic assemblages may be better explained by these implicit operational variables than by the mapped variables, but the mapped variables are used because they provide a means of recognizing and delineating units on the earth's surface. The mapped variables are assumed to be adequate surrogates for many other variables. Slope and aspect, for instance, provide information about insolation and water relationships and hence about productivity. The use of geocomplexes does not rest on being able to specify these relationships. It rests on an empirical understanding of the connection between abiotic components and their relation to biota, so that the mapping of simple abiotic features provides an operational basis for making statements about ecosystem functions.

A further aspect of these units is that neighbouring biogeocoenoses are arranged in specific relation to each other. For instance, in valleys, river meads, slopes, terraces and channels have recognizable chorological (spatial) structures. In hilly areas, slopes, interfluves, terraces, mass movement features and deposits have recognizable, mappable spatial relations. Adjacent geocomplexes affect each other through fluxes of matter and energy so that the condition of one geocomplex controls another and is in turn controlled by it. These relationships, depending in large part on gravity-driven movement, produce chorological structures which mirror drainage patterns.

The mapping of geobiocoenoses may appear to be a rather naive approach to vegetation because of the rather obvious point that 'natural' vegetation patterns are unmappable. They are not simple structures controlled by abiotic factors. The response of vegetation to abiotic factors can be described only in the most general terms. Our knowledge of vegetation behaviour and population and community dynamics has a very limited empirical base (Harper, 1985). However, such reductionist arguments ignore two points about landscape ecology. First, it operates at map scales between about 1 : 5000 and 1 : 1 000 000—in other words far above the scale of the size of or area of influence of an individual plant. At these scales relationships between biotic and abiotic characteristics of areas are manifest and interpretable. Secondly, no part of the European landscape is natural. Even naturally regenerating areas are heavily influenced by the hand of man. Those areas and sites which come near to the natural state are generally protected to some degree or other, i.e. managed.

This means that vegetation exists as blocks and patches which are intrinsilcty mappable. The natural variabilty in biota and the complexity of the relationships between biotic and abiotic factors have been removed or masked. It is appropriate in such landscapes to use units as if they were homogeneous and discrete. Individual geocomplexes can be considered as sites and their chorological structures as real patterns in nature.

The practice of landscape ecology is about the creation of a balanced, but also dynamic, rational relationship between the natural properties of an area, its geo-ecological structure and its economic use. Landscape ecological management aims to (Bucek and Lacina, 1985):

1. ensure favourable effects of land use and management in an area on adjacent ecologically less stable parts of the landscape;
2. preserve the landscape geofond;
3. preserve valued landscape features; and
4. promote a polyfunctional use of the landscape.

There are many systems and terms used for landscape ecological studies. The land-unit and land-system ideas of the CSIRO (Christian and Stewart, 1968) and the system of ecotope, land-facet, land system and main landscape of Zonneveld (1972) are two examples.

Perez-Trejo in Chapter 7 of this volume provides similar ideas about landscape response units. He is concerned with justifying the use of units by considering how they have evolved. In Chapter 6, Cousins deals with another aspect of units within landscapes, the existence of a hierarchy. Both these ideas are important parts of the conceptual basis and operational approach of landscape ecology. However, most systems and approaches presented in western literature express only a limited idea of the relationships in the landscape which need to be considered in an operational landscape ecological approach to landscape management.

The relationships in the landscape have topological, chronological and choro-logical dimensions. Topological relationships between the structures are established by evaluating the most appropriate types of economic and ecological use of geo-complexes for the given type of natural properties (Hasse *et al.*, 1986). Chronological relationships are about the continuous process of landscape management. Chronology provides both the means of understanding landscape development and of detecting causal relations in the landscape. Thus it is a fundamental part of making prognoses for landscape planning (Kolejka, 1988). Data on geocomplexes arranged as maps in chronological order serve as a starting point for evaluation and decision-making in planning. The task of the landscape ecologist is to answer two questions. What types of economic function and changes in economic function characterize a given type of geocomplex? How do the spatial relationships between geocomplexes affect the dynamics of the utilization of areas? The analysis involves determining the character of the functional stability and dynamics of geocomplexes and marking out critical sites and zones of conflict.

Chorological relationships require the construction of an optimal functional structure in the landscape based on the geoecological structure, but enabling poly-functional use of the geocomplexes. This territorial concept of ecological stability is an important part of designing optimum use of landscapes. The method assumes that if highly productive use (or some potentially disruptive use) is to be maintained on some landscape elements, and the cultural landscape as a whole is to be main-tained, then it is necessary to isolate ecologically unstable parts by a system of stable and stabilizing ecosystems at both local and regional levels.

Bucek and coworkers have designed skeletons of ecological stability at the national scale with major corridors which connect biocentres of national and inter-national importance forming broad regional systems, where stabilizing uses should predominate. On larger scales centres of regional and local importance are con-nected by land parcels with stabilizing uses.

The design of skeletons of stability follows a defined procedure (Bucek and Lacina, 1979, 1981), i.e. division of the landscape into areas with specific evological conditions, termed 'geobiocoenes'. These correspond to the geocomplexes described above.

1. Assessment of the current state of vegetation in these areas.
2. Statements of the functional possibilities of each area and the significance of spatial groupings of biogeocoene types.
3. Characterization of geobiocoenoses according to the intensity of economic activity.
4. Definition of ecologically important landscape elements.

Landscape ecology can be seen, therefore, to encompass ideas not only about natural landscapes but also about economic landscapes. Indeed the landscape is seen to be quintessentially the product of economic activity over long periods of time. The economic dimension manifests itself as a mosaic of functionally differen-tiated areas within a territory. The degree of correspondence between the geo-ecological and the economic functions of the landscape is a consequence of the cultural and social development of the people of an area, their level of knowledge about the environment, and their technological and organizational skills.

Current landscape studies in Czechoslovakia

A balanced or exhaustive review of Czechoslovak landscape ecology is not possible here. The variety of approaches to landscape study is thus perhaps best illustrated by considering a selection of problems and topics which have been dealt with on different scales. In the work of the Academies of Science, studies have been planned and executed at what can be considered to be three broad levels: national, regional

and local. The distinction between levels may be arbitrary, but there are real differences in the problems and in the types of data which are analysed on the different
scales. Of more importance is that studies at the regional and local levels are undertaken in the context of the national systems of landscape. There is a conscious
hierarchical structure in landscape studies, especially in terms of ecological stability.

National studies

There has been a series of studies of the entire Czechoslovakian Republic published
at scales of 1 : 500 000 and 1 : 1 000 000 (Pribyl and Stehlik, 1988). A good example
is the map of the environment of the CSSR made by Ivan *et al.* (1987) published at a
scale of 1 : 750 000 (Figure 4.1). It presents a complete synthesis of grades of areas
and sites with polluted and damaged land, water and air. It shows the major zones
of environmental quality at a national scale and is meant to be used as a tool for
prioritizing direct action and taking preventive measures as well as being a broad
guide to the state of the environment.

   An example of a skeleton of ecological stability at the national scale is provided
by Miklos (1988). This shows the major biocentres of different levels of importance
in Slovakia (Figure 4.2) with the suggested linking zones or corridors of ecologically

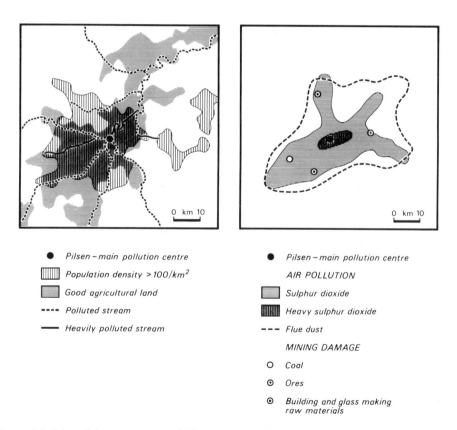

● Pilsen – main pollution centre	● Pilsen – main pollution centre
▦ Population density > 100/km²	AIR POLLUTION
▦ Good agricultural land	▦ Sulphur dioxide
---- Polluted stream	▦ Heavy sulphur dioxide
—— Heavily polluted stream	--- Flue dust
	MINING DAMAGE
	○ Coal
	◉ Ores
	◉ Building and glass making raw materials

*Figure 4.1. Map of the environment of Plzen area, Czechoslovakia based on detail of the Map of
the Environment of the Atlas of Population of CSFR (Ivan et al., 1987). In the original all data
is plotted on a single sheet using colour. Here black and white necessitates separation of data.*

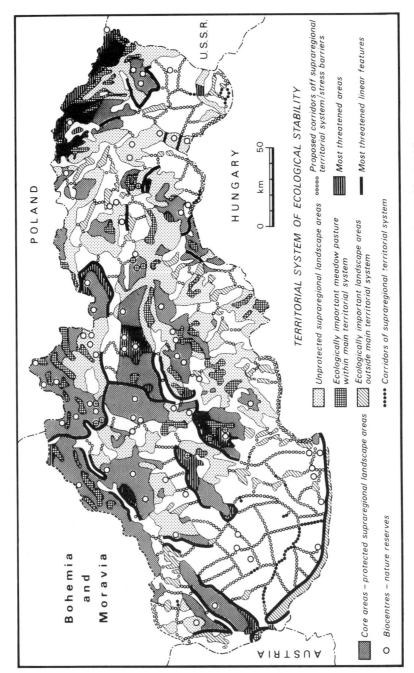

Figure 4.2. A skeleton of ecological stability of Slovakia based on Miklos (1988).

more stable ecosystem types. Together these are planned to form a connected system of stabilizing areas isolating those areas subject to stress and instability. Such maps are partly statements of the actual state of the environment and partly planning statements, because considerable areas of land need to be designated and controlled before the systems are complete. When combined with maps of ecological stress (Figure 4.3) it forms an important management tool showing which areas need priority attention.

The general ecological map of the Czech lands produced by the Czechoslovak Academy of Sciences (CSAV) in 1985 at a scale of 1 : 500 000 provides an analysis of the states of individual parts of the environment considered in terms of ecological stress. The map units are based on the ecological theory of stress from biotic and abiotic factors. It is meant to indicate at a strategic level where symptoms of stress are likely to be found, and hence to provide a means of assessing areas of damage and conflict in the environment. It is also a first basis for monitoring.

These small-scale maps with their reports show clearly the approach to landscape study at the national level and give us an appreciation of the extent to which whole landscape thinking permeates environmental work in central Europe. It cannot be overemphasized that such awareness of landscape structures and character at the national level and at larger scales is an almost unconscious tradition of both applied and basic scientific work.

Regional studies

For regions and districts, landscape studies are carried out at scales of 1 : 100 000 to 1 : 25 000. It is at these and larger scales that exercises of 'geographic differentiation' are undertaken in which land units, or geocomplexes, are defined and delineated. These are based on systematic surveys of topography, geology, soils, climate and so on. As examples, the geographic differentiation of the Southern Moravian region (Kolejka and Kolejkova, 1988) and the Blansko region (Lacina and Quitt, 1986) are presented at a scale of 1 : 100 000, but are based on field work and preliminary mapping at scales of 1 : 50 000 and locally at 1 : 10 000. Such maps of geographical units are intended to form the basis for land planning and for analyses of ecological stability.

The Slovak Academy of Science has developed a more applied approach to land planning in its landscape ecological planning (LANDEP) method (Ruzicka and Miklos, 1982). It uses essentially the same approach but has been developed to a much greater degree as a management tool. There are two special points of emphasis. First, the algorithms for determining the operational characteristics of the predetermined geocomplexes form basic survey data in relation to particular objectives. So, for example, in a study of erosion control on farmland in eastern Slovakia (Ruzicka and Miklos, 1986) the data on topography, soil, hydrology and so on are combined in one way to derive descriptors of the geocomplex related to erodibility. In studies of production potential in the same area, some or all of the same data types and other data are combined in a different manner to give appropriate measures of productivity. The second point of emphasis is in the method of allocating land uses or practices to geocomplexes. This involves a process of assigning measures of suitability of all uses to all types of geocomplex from ideal to damaging/unsuitable. The allocation of uses to units can then be optimized within other operating parameters such as the economic mix of land uses in an area or the skeleton of ecological stability.

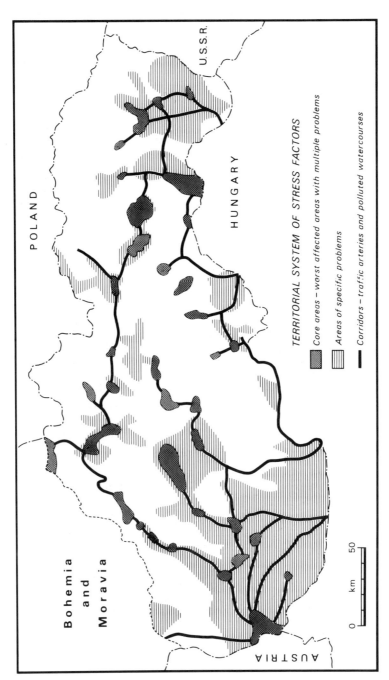

Figure 4.3. A map of ecological stress zones on Slovakia based on Miklos (1988).

A regional-scale example of the integration of methodology of landscape ecology into a GIS environment is given by the joint programme of the Geography Institute of the Czech Academy of Science, the Czech Institute for Nature Protection and the University of Salford, UK. This work is developing a management information system for the Zdarske vrchy landscape protected area in the Bohemia–Moravia highland. It uses maps of the skeleton of ecological stability derived from classification of LANDSAT TM data figure 4.4 (Pauknerova and Brokeš, 1992) with a map of biogeocoenoses produced by conventional survey and mapping (Bucek and Lacina, 1975) (Figure 4.4). One of the regional-scale problems addressed by the GIS is the modelling of future damage impacts of SO_2 on forest. The modelling uses current land-cover information derived from LANDSAT data on spruce and other forest species together with topographical data and the map of biogeocoenoses. Based on empirical findings, the model assumes that damage occurs predominantly above 700 m, predominantly on spruce species and on particular biogeocoenoses. The biogeocoenoses act as a basis by providing a complex of information on natural conditions such as soil, microclimate and hydrology. For each biogeocoenosis type it is possible to speculate whether it is suitable for spruce and/or other forest species, and whether in relative terms the species are susceptible to damage under the conditions prevailing there. From this information on altitude, species and biogeo-

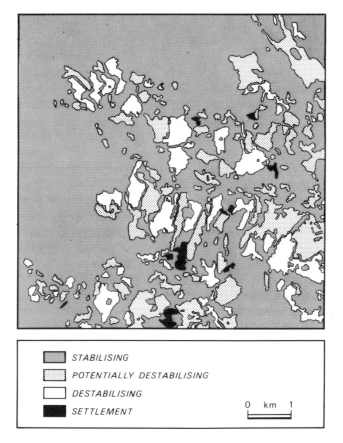

STABILISING

POTENTIALLY DESTABILISING

DESTABILISING

SETTLEMENT

0   km   1

Figure 4.4. Map of ecological stability of part of the Zdarske vrchy protected area using land cover data from LANDSAT TM and a map of biogeocoenoses (Buček and Lacina, 1975).

coenoses it is possible to build a simple landscape model for worst-case, best-case and other scenarios (Bucek *et al.*, 1990). These maps can then be used in planning responses to forest damage, not only in terms of loss of forest resource, but also in terms of changes in landscape stability because of the loss of stabilizing land cover as expressed in the skeleton of ecological stability.

Another study is of the black grouse (*Tetraotetrix*) (Brokes *et al.*, 1991), which has been in decline for many years because of the changing composition of the landscape following land improvement and deforestation. The first objective of this study is to identify bird habitats using contemporary and historical data on distributions of birds and land cover. The bird in its habitat is known to need a particular mix of land-cover types and conditions. These can be determined from satellite data and physiographic maps. These habitat models can be translated into landscape models of potential bird areas. However, the preference of the bird for particular land-cover conditions depends on the quality of that cover, which is determined by abiotic conditions, as expressed in biogeocoenoses. Empirical data from current and past distributions of the bird show a preference for a very limited number of biogeocoenoses, irrespective of vegetation cover. To some extent this can be understood in terms of feeding habits of the bird and the species in hedgerows, in unmanaged areas and in the understorey of forest and woods. The species in these communities are unlikely to change as much as dominant woody species, and thus will reflect the mapped biogeocoenoses even in managed areas. Further work will aim to identify sites for restocking of the bird and to model impacts of possible land-use changes through use of skeletons of ecological stability.

Local studies

At the local level, using map scales of 1 : 50 000 to 1 : 10 000, landscape studies may deal with planning land use on individual farms or the impact of specific technical projects. A nice example of large scale ecological planning is Rozova's (1988) study of verdue in rural settlements. She uses data on a scale of 1 : 10 000 on functional and spatial categories of land in villages in Slovakia. These are used first to assess settlement stucture and then to evaluate the role and relationships of the biotic components of the structure to other components by examining their spatial relationships and possible conflicts. Finally, recommendations are made for preserving or changing the biotic cover so that there is a compatibility between biotic and non-biotic components and the verdue structure is stable. Her analysis takes account of both economic and social characteristics of land and of the suitability of parcels for different uses based on abiotic characteristics. The result is a map of proposals for verdue in a settlement (Figure 4.5) which includes recommendations on establishing new areas and improving existing ones so that each part of the village can operate efficiently.

Landscape ecology and GIS

The remarkable rate of technical development in the field of GIS has brought into focus the issue of whether or not technological developments are conceptually and scientifically appropriate for solving scientific problems. The capacity of modern computers to handle and display spatial data is no guarantee that what is being

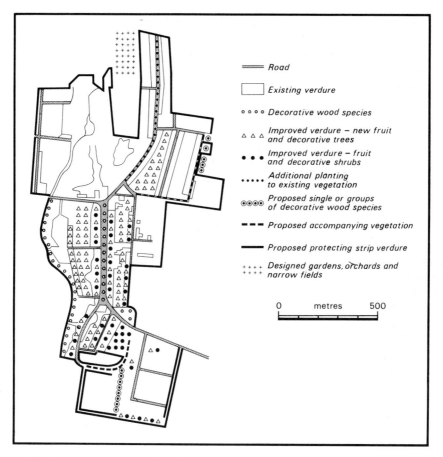

Legend:

═══ *Road*

☐ *Existing verdure*

∘ ∘ ∘ ∘ *Decorative wood species*

△ △ △ *Improved verdure – new fruit and decorative trees*

● ● ● *Improved verdure – fruit and decorative shrubs*

∙∙∙∙∙ *Additional planting to existing vegetation*

⊙⊙⊙⊙ *Proposed single or groups of decorative wood species*

▬ ▬ ▬ *Proposed accompanying vegetation*

▬▬▬ *Proposed protecting strip verdure*

+ + + +
+ + + + *Designed gardens, orchards and narrow fields*

0   metres   500

Figure 4.5. Landscape structure of verdue in a Carpathian village based on Rozova (1988).

displayed is providing the answer that is needed to any particular problem. Indeed, there is a great danger that the power of the digital display may divert attention away from the central issue of pursuing scientific environmental analysis. For large-area studies in particular, there are dangers arising from the apparent veracity of data for areas which have not been measured, or of interpolated data. From first principles we can argue that much of the character of an area is related to and can be deduced from basic geographical parameters such as topography, climate and geology. The best we can do is to use these data in our analyses of the environment and to use data in a way that we can understand and control. We need to deal with parcels, units or geocomplexes of land. First principles also tell us that the spatial relationships, or chronological structure, of things are important in determining their behaviour. Practice at thinking about and working with land units and their spatial relationships is rather rare in Anglo-American landscape studies, but not in central European work. The point here is that we probably have something to learn from studying the methods of landscape analysis developed over the last 50 years in central Europe, which could stimulate and direct GIS work in the West.

Acknowledgements

The authors thank Gustav Dobrynski for drawing the diagrams.

References

Brokeš, P., Downey, I., Heywood, I., Pauknerova, E. and Petch, J., 1991, Remote Sensing and GIS in Management of Protected Areas. IUCN Conference of Science and Management of Protected Areas, Acadia, Canada: Wolfville University.

Buček, A. and Lacina, J., 1975, Biogeograficka Characteristica CHKO Zdarske vrchy, Geograficky Ustav, Brno: CSAV.

Buček, A. and Lacina, J., 1979, Biogeograficka diferenciace krajiny jako jeden z ekologickych podkladu pro uzemni planovani, *Uzemni planovani a urbanismus*, **6**(6), 382–387.

Buček, A. and Lacina, J., 1981, The use of biogeographical differentiation in landscape protection and design, *Sbornik CSGS, Praha*, **86**(1), 4–50.

Buček, A. and Lacina, J., 1985, The skeleton of ecological stability in landscape planning, *International Symposium on Emerging Regional Ecological Problems*, Brno: Czechoslovak Academy of Science.

Buček, A. and Mikulik, O., 1977, Valuation of the negative effects of economic activities on the environment of the model region of Liberec and excursion guide, *Studia Geographica 57*, Brno: Geograficky ustav CSAV.

Buček, A. and Mikulik, O., 1988, The methodology of geography and environment and its application for regional research, *Sbornik praci, CSAV*, **15**, 151–157.

Buček, A., Downey, I., Pauknerova, E., Petch, J. and Petrlik, J., 1990, A GIS approach to predicting forest damage. IUCN Conference on Parks, People and Pollution, Czechoslovakia: Krknose.

Bunting, B., 1989, The turning of the worm: early nineteenth century concepts of soil in Britain—the development of ideas and ideas of development, 1834–1843, in Tinkler, K. J. (Ed.), *History of Geomorphology (Binghampton Symposia in Geomorphology No. 19)*, pp. 85–197, London: Unwin Hyman.

Christian, C. S. and Stewart, G. A., 1968, Methodology of integrated surveys, in *Aerial Surveys and Integrated Studies. Proc. Toulouse Conference*, pp. 233–280, Paris: UNESCO.

Colinvaux, P., 1973, *Introducton to Ecology*, New York: Wiley.

Dedina, V., 1927, Prirozene kraje a oblasti v Ceskoslovensku, *Sbornik CSSZ*, **30**(1), 21–25.

Dvorsky, V., 1918, Uzemi ceskeho naroda, *Cesky ctenar*, **10**(6–7), 1–79.

Forman, R. T. T. and Godron, M., 1986, *Landscape Ecology*, New York: Wiley.

Gardavsky, V., 1988, To the prognosis of geography. Institute of Geography of the Czechoslavak Academy of Sciences, *Brno. Sbornik Praci*, **15**, 207–220.

Goodchild, M. F., 1990, Spatial information science. Keynote address in *4th International Symposium on Spatial Data Handling, Zurich*, Vol. 1, pp. 3–12.

Haines-Young, R. and Petch, J. R., 1985, *Physical Geography: Its Nature and Methods*, London: Harper and Row.

Harper, J., 1985, *Population Biology of Plants*, London: Academic Press.

Hasse, G., Mannsfeld, K. and Schmidt, R., 1986, Types des Anordnungsmusters zur Kennesochnung der Arealstruktur von microgeochoren, *Petermans Geografisk Mittelungen 1, Gotha/Leipzig*, **86**, 31–39.

Hradek, M., 1988, Some examples of applied geomorphological maps from Czechoslovakia, *Zeitschrift fur Geomorphologie* N.F., Supplementband, Berlin/Stuttgart: Gebruder Borntrager.

Hribova, B., 1956, Mapa prirodni krajiny Ceskych zemi ve 12. stoleti (mapa a pruvodni text), in *Sbornik Vysoke skoly pedagogicke, Prirodni vedy II*, pp. 61–94, Praha: SPN.

Hromadka, J., 1943, Vseobecny zemepis Slovenska, in *Slovenska vlastiveda I*, pp. 81–332, Bratislava: Slovenska akademia vied a umeni.

Hynek, A., 1981, Integrated landscape research, *Scripta Faculta Scienta Universita Purkyneusis, Brno*, **7–8**, 283–300.

Ivan, A., Kriz, H., Lacina, J., Mikvlik, O., Quitt, E., Ungerman, J., Vaishar, A., Vlcek, V. and Zapletalova, J., 1987, Map of Environment, Population Atlas of the CSSR, Brno: Geograficky Ustav CSAV.

Kolacek, F., 1924, Prirodni krajiny na Morave a v ceskem Slezsku, in *Priroda 17*, pp. 249–253, 314–325, Brno: Barvic a Novotny.

Kolejka, J., 1988, Development of progressive traditions of the complex research of landscape in Czech geography in the Institute of Geography of the CSAV, *Brno Sbornik Praci* **15**, 123–150, CSAV.

Kolejka, J. and Kolejkova, D., 1988, 'Geoekologika mapa Jihomoravsk kraje', Geograficky Ustav, CSAV, unpublished manuscript.

Korcak, 1936, Regioalni typ v pojeti statistickem, in *Sbornik III. sjjezdu ceskoslovenskych geografu v Plzni*, pp. 111–112, Praha: Ceskoslovenska graficka unie A.S.

Kral, J., 1930, Uvahy o rozdeleni ceskoslovenskych Karpat na prirodni oblasti a pojmenovani techto oblasti, *Sbornik filosoficke fakulty University Komenskeho*, **7**, No. 54(1), 1–33.

Lacina, J. and Quitt, E., 1986, Geografika diferenciace okresv Blansko, in *Geografie: teorie-vyzkim-praxe*, Vol. 3, pp. 1–210, Brno: CSAV.

Mikos, L., 1988, Space and Position—scene of the origin of spatial ecological landscape problems, in *Proc. VIII International Symposium Problems of Landscape Ecology*, Vol. 1, pp. 51–72, Bratislava: Slovak Academy of Science.

Murdoch, W. W., 1966, Community structure, population control and competition—a critique, *The American Naturalist*, **100**, 219–226.

Naveh, Z., 1982, Landscape ecology as an emerging branch of human ecosystem science, *Advances in Ecological Research*, **12**, 189–247.

Naveh, Z. and Lieberman, A. S., 1986, *Landscape Ecology: Theory and Applications*, New York: Springer-Verlag.

Odum, E. P., 1969, The strategy of ecosystem development, *Science*, **164**, 262–270.

Pelisek, J. and Sekaninova, D., 1975, Pedologicka regionalizace CSR, *Studia Geographica 49*, Brno: Geograficky ustav CSAV.

Pribyl, J. and Stehlik, D., 1988, History and the most important results of the Institute of Geography of the Czechoslovak Academy of Sciences (CSAV), *Sbornik Praci (CSAV)*, **15**, 7–22.

Risser, P. G., 1985, Towards a holistic management perspective, *Biological Sciences*, **35**, 414–418.

Risser, P. G., 1987, Landscape ecology: state-of-the-art, in Turner, M. G. (Ed.), *Landscape Heterogeneity and Disturbance*, New York: Springer-Verlag.

Risser, P. G., Karr, J. R. and Forman, R. T. T., 1984, Landscape Ecology: Directions and Approaches, *Illinois Natural History Survey Special Publication No. 2*.

Roughgarden, J., May, R. M. and Levin, S. A. (Eds), 1989, *Perspectives in Ecological Theory*, Princeton, NJ: Princeton University Press.

Rozova, Z., 1988, Verdue structure in a rural settlement, in *Proc. VIII International Symposium on Problems of Landscape Ecological Research, IALE, Zemplinska Sirava*, Vol. 1, pp. 324–348, Bratislava: Slovak Academy of Science.

Ruzicka, M. and Miklos, L., 1982, Landscape-ecological planning (LANDEP) in the process of territorial planning, *Bratislava, Ekologica (CSSR)*, **1**(3), 297–312.

Ruzicka, M. and Micklos, L., 1986, *Basic Premises and Methods in Landscape Ecological Planning (Modelling)*, Bratislave: Slovenska Academic Ved.

Schum, S. A. and Lichty, R. W., 1965, Time space and causality in geomorphology, *American Journal of Science*, **263**, 110–120.

Tinkler, K. J. (Ed.), 1989, History of geomorphology, *Binghampton Symposia on Geomorphology*, No. 19, pp. 85–107, Unwin Hyman.

Turner, M. G. (Ed.), 1987, *Landscape Heterogeneity and Disturbance*, New York: Springer-Verlag.

Vitoussek, P. M. and Reiners, W. A., 1977, Ecosystem succession and nutrient retention: a hypothesis, *Bioscience*, **26**, 376–381.

West, D. C., Shugart, H. H. and Botkin, D. B. (Eds), 1980, *Forest Succession, Concepts and Applications*, New York: Springer-Verlag.

Whittaker, R. H., 1975, *Communities and Ecosystems*, 2nd Edn, New York: Macmillan.

Zonneveld, I. S., 1972, *Textbook of Photo-Interpretation*, Enschede: ITC.

5

Equilibrium landscapes and their aftermath:
spatial heterogeneity and the role of new
technology

C. P. Lavers and R. Haines-Young

Introduction

Landscape ecology deals with the structure of landscapes, the way they function and
the way they change. As Forman and Godron (1986) point out, since we have
always lived in landscapes, the roots of the discipline can be traced back to some of
the earliest scientific writings. As a distinct discipline, however, it was not until the
post-World War II period that it emerged as an area of study in its own right. How
did such a research focus develop? What insights does it offer compared with other
areas of environmental science?

The problem one faces in answering such questions about the rise of landscape
ecology is that, despite its short history, it has already supported several distinctive
research themes. To many, for example those working in the Central European
tradition, the subject serves mainly to unify our understanding of environmental
systems. The landscape concept is used to stress interrelationships between the
physical, biological and cultural aspects of ecological systems. Much effort is
directed towards mapping and describing landscape patterns, with the aim of using
these data for planning purposes.

By contrast, other workers have used the landscape concept to emphasize quite
different concerns. Many landscape ecologists in North America, and some in
Western Europe have focused primarily on the factors which control the location
and action of organisms in space and time, and on the influence which organisms
have on landscape pattern. The approach here, compared with that of the Central
European tradition, is analytical rather than descriptive. The aim is to understand
the spatial structure of ecosystems rather than to document the properties of landscapes as a whole.

The characteristics of the Central European tradition are described elsewhere in
this volume (see Chapter 4). In this chapter we reflect on this other body of work
which emphasizes the interactions between organisms and landscape. Our interest
comes about because, although the study of the influence of landscape factors on
ecological systems has long been the concern of ecologists and biogeographers,
landscape ecology is now often represented as providing a somewhat new perspective on these old problems. The purpose of this chapter is to examine this claim
critically.

The development of the 'landscape ecological approach' is first considered in relation to the ideas which have formed the core of much ecological research over the last three decades. It is shown that the 'new paradigm' does contain elements which distinguish it from what has gone before, but that these ideas remain largely untested. This chapter concludes with a review of the role which spatial information systems might play in helping to refine current concepts. The case studies we have selected mainly reflect our interest in avian ecology.

MacArthur and the 'conventional wisdom'

Many of the ideas that have dominated ecology during the last 30 years had their origins in avian community ecology. The work of Robert MacArthur and his colleagues (MacArthur, 1957, 1958, 1965, 1970, 1972; MacArthur and MacArthur, 1961; MacArthur and Levins, 1967; MacArthur and Wilson, 1967) did more than just explain ecological patterns. The system of investigation which developed from their work, using competition as the main determinant of community structure, became central to much ecological thinking. So dominant was the tradition that these studies represented that for some authors, the 'MacArthur approach' exhibited many of the attributes of a 'Kuhnian paradigm' (Wiens, 1989: Vol. 1; Kuhn, 1970) that is, some dominant world view which serves to unify a discipline.

Wiens, for example, writes:

> It [the 'competitive paradigm'] presented a coherent unified tradition of scientific research that defined which problems were interesting and merited study and which were not . . . And it discouraged alternative explanations or views . . . (Wiens, 1989: Vol. 1, 10–11)

Although some would dispute the overriding dominance of MacArthur's influence, it is clear that for many the ideas held centre stage. Price *et al.* (1984), for example, suggest that researching ecologists and journal editors were quick to publish evidence supporting the central ideas of the paradigm, however misleading, while '. . . a scientifically rigorous demonstration of competition's absence in a certain community [was] regarded as uninteresting'. Gilbert (1980) describes a similar situation in the context of the development and testing of MacArthur and Wilson's theory of island biogeography.

It is unfortunate that a 'paradigm' has come to bear MacArthur's name. For, although he may have popularized its central ideas, the 'intellectual tyranny of the concept of competition and . . . the continuing prevalence of deterministic, equilibrium models in ecological theory' (Colwell, 1984), required a coordinated effort by many ecologists. A dogmatic adherence to any theory is not a trait which characterizes MacArthur's approach to ecology. He was always critical of his own ideas. It has only been much later that ecologists more generally have adopted a similar attitude, recognized the limitations which the paradigm imposed and sought a new focus for their work. One such development was the new emphasis on the analysis of landscape structure.

What aspects of the 'competitive' paradigm are particularly restrictive compared to the approach now proffered by landscape ecologists? Two ideas central to the paradigm can be identified, namely:

1. that communities are at or close to equilibrium and that interspecific competition is the main determinant of community structure; and

2. that the search for general and repeated patterns is of paramount importance.

We review each of these in turn and show how landscape ecology attempts to move
beyond the 'conventional wisdom' embodied in these ideas.

Equilibrium and competition

The 'competitive paradigm' can be viewed as restrictive in that it focused prefer-
entially on 'equilibrium communities' in which competition was the major force
promoting structure. Indeed, so dominant were these assumptions that for many,
only communities at or near to equilibrium were legitimate objects of study. The
basis of this assertion was the assumption that the causal link between the factors
promoting community structure, such as competition and the resulting community
patterns, would be complicated if individuals were responding to external disturbance. The obvious criticism here is that in equilibrium communities, the caus-
ally efficacious mechanism promoting community structure can only be inferred
from the resulting pattern. In a perfectly equilibrial community, all that remains of
the structuring *process* is the stabilizing selection which maintains the pattern.
Beyond this, there is also the question as to what extent community equilibrium is
actually normal, and competition a major structuring force.

Wiens (1984), reporting on studies of shrub-steppe birds in western North
American landscapes, relates how these birds:

> ... seem not to be in any readily definable equilibrium, vary pretty much inde-
> pendently of one another, do not appear to be resource limited very often, do
> not fully saturate the available habitat, and do not provide clear evidence of
> ongoing competition. We seem to be dealing with non-equilibrium, non-
> competitive communities by any conventional definition of these terms.

The studies of Wiens and colleagues (e.g. Rotenberry and Wiens, 1980a, b; Wiens,
1986, 1987), concentrate on arid and semi-arid grasslands and shrub-steppe landscapes in North America. But with similar landscapes comprising 34% of the land
area of the earth (40% including tundra (Wiens, 1984: 450)) it is questionable
whether there is any justification for assuming the 'ubiquity' of community equilibrium.

The assumption of community equilibrium requires the additional assumption
that organisms rapidly track changing environmental conditions. This in itself limits
the scale of investigation to individuals or local populations where time lags in
response to change may be minutes to a few years, and generally would exclude
consideration of dynamic processes on a regional scale where time lags may be of
the order of years to decades or longer (see Wiens, 1989: Vol. 2, table 5.3).

Some species respond rapidly to environmental change, but some do not. The
association of two such species in a community may say more about their differing
sensitivities to change than about competitive equilibrium. An excellent review of
the effects of time lags in response to climatic change is given by Davis (1986). For
example, in the UK a distinct amelioration of climate occurred between *c.* 1890 and
1950, followed by a cooling from 1950 to the present day. Williamson (1975) shows
that at least three species of bird (great spotted woodpecker, green woodpecker and
starling) were still expanding northward into Scotland in 1975 in response to the
pre-1950 amelioration, a time lag in response of 25 years. Simultaneously, osprey

and snowy owl—species which track climatic changes more closely—have been migrating southward in response to the post-1950 cooling. Wiens (1989: Vol. 1, Chap. 8) presents many other examples of how climatic and physiological tolerances of bird species determine their distributional ranges.

Not only may organisms be slow to track environmental change by colonization, but once occupying an area, they may linger long after conditions are no longer optimal. Wiens (1986) relates how sage sparrow densities showed a good deal of inertia, remaining unchanged after their habitat had been drastically altered by spraying of the study plot with herbicide. Similarly, no clear response to the habitat alteration was shown by Brewer's sparrow in the years after spraying. It is thought that the lack of immediate response is due to fidelity of males to previously occupied territories.

So, is the existence of community equilibrium a valid assumption? Under certain circumstances, probably; but these circumstances are necessarily restrictive. Even in single habitat patches (although it is equally applicable to larger areas), long-term processes such as climatic change, and differential rate of response by organisms, may produce community patterns which are not explicable in terms of proximal habitat features and competitive interactions. Broadening the scale beyond that of the equilibrial habitat patch, processes of disturbance come more into play, with consequences for competitive assortment and the attainment of equilibrium. With disturbance regimes operating at rates faster than the rate of competitive exclusion, and with disturbance patches accessible to poorer competitors, there may be a permanent (although spatially transient) reservoir of poorer competitors in a given region.

Although the assumptions about equilibrium and competition are difficult to disentangle within the conventional wisdom, it does not follow that in equilibrium communities competitive interactions are the major force promoting community structure. This is a quite separate assumption which needs to be considered.

Ecological pattern has often been ascribed mainly to the operation of interspecific competition by a number of workers (e.g. Diamond, 1978; Cody, 1981). Criticism of this assertion are many and varied. We will focus on just one line of attack, namely the debate centred on the question; 'Under what environmental circumstances is competition likely to be the major force shaping community structure?'.

Price (1984) suggests that, when considering the circumstances under which competition is likely to occur, the landscape scale is generally not one at which such interactions are likely to be seen. Landscapes are spatially heterogeneous in their structure along many resource axes. The extent to which organisms enter into competition will depend not only on the spatial and temporal availability of resources, but also on characteristics of the organism concerned.

For example, herbivores feeding on rapidly increasing deciduous leaf resources in temperate regions are less likely to enter into competitive interactions if the generation cycle is annual, than if the cycle runs to several generations per year (Price, 1984: 359–360). Alternatively, if resources are distributed in a patchy manner, then the probability of colonization of such patches by potential competitors is reduced and competition, as a process of controlling community structure, may not be significant.

Equilibrium communities constructed by competitive interactions are likely to be highly indicative of stable environments. However, as stable environments probably exist only on a very restricted range of spatial scales some alternative world view may be required. The more frequent condition is that resources are patchy and disturbance is likely to disrupt biotic interactions. Therefore, in contrast to what has gone before, landscape ecology gives greater emphasis to notions of disequilibrium and the exploitation of spatially heterogeneous and ephemeral resources.

When the assumptions implicit in the 'competitive paradigm' are considered, it is hardly surprising that resource patterns at the landscape scale were considered as 'uninteresting' by a research paradigm which focused mainly on equilibrium communities. Equally, the attention given to spatial heterogeneity by the 'new' landscape ecology serves to distinguish it from much of what has gone before.

Generality and the role of theory

In addition to scientific assumptions about equilibrium and competition, the 'conventional wisdom' also contained several key methodological elements which did much to shape work which followed in its tradition. An underlying goal of the 'competitive paradigm', for example, was the emphasis it gave to the search for general, that is repeated, patterns of community structure.

Ignoring the complexity of individual behaviour in patchy environments, the competitive paradigm aimed particularly at the mathematical descriptions of ecological pattern. These descriptions were constructed in terms of highly predictable, deterministic processes, processes which are relevant to only a narrow range of circumstances. As we have seen, the systems chosen for description tended either to be undisturbed communities exhibiting simple community patterns, or larger scale, but nevertheless simple systems such as island communities. The corollary of the approach, of course, was that systems which failed to elucidate clear and repeated patterns were considered uninteresting. If this point is accepted, it is perhaps not surprising that the landscape did not emerge as a useful scientific concept until the dogma of the competitive paradigm had been removed, and its monopoly of the ecological literature ended.

Recognition that environments are spatially and temporally heterogeneous, that no two areas have the same spatial arrangement of resources, or the same history, brought into question whether general principles about equilibrium community structure at any scale greater than the undisturbed patch, could even be identified. Landscapes are complex, dynamic entities, and, when studying organisms in the landscape details, particularly historical details, matter.

The contrast between the ideas of equilibrium and uniformity, which were such important parts of the 'competitive paradigm', and the notions of individuality, heterogeneity, and unpredictable change, given emphasis by landscape ecology, could be viewed as equivalent to the distinction between ahistorical and historical perspectives described by Kingsland (1985). Noting the methodological implications of the ideas, she argues that the incorporation of historical factors, such as time lags, into mathematical models of community structure soon resulted in mathematical intractability. Faced with the complexity of real, non-equilibrium landscapes, it would seem that the competitive paradigm had nowhere to go.

The extent to which the rise of landscape ecology marks the growth of a fundamentally new paradigm is contentious. Although some would prefer to present the situation in these stark terms, there are others who emphasize the complexity of the situation. Colwell (1984), for example, interprets the reformulation of ideas in mainstream ecology not as a 'paradigm shift' as presented here, but as a 'salubrious readjustment in the balance between our increasingly detailed appreciation of nature, and the domain of our theories and models'. He goes on:

> The special features (adaptive or not) of particular kinds of organisms constrain the ecological generalities they can support. In the other direction, generalities define the expected and thereby permit us to identify and explore the exceptional. When the balance between these two processes swings too far in one direction or the other, the explanatory power of a science suffers: a compendium of case histories is unenlightening . . . and a generality is of little use if the exceptions far outweigh the cases that approximate the expected. (Colwell, 1984: 393–394).

Whatever the case, the most important thing to note is that the landscape represents an entity of greater complexity than systems which can usefully be used to investigate the ideas of the competitive paradigm. This does not imply that general principles about landscape function and communities structure cannot be recognized, or that these principles will not be amenable to theoretical explanation. In the present context, the point being made is that a certain class of models, represented by the competitive paradigm, are limited in their usefulness, and that alternative approaches are required. The extent to which landscape ecology can provide new insight remains to be seen.

The influence of landscape structure

Landscape ecology emphasizes the mosaic structure of landscapes and the influence of spatial heterogeneity in ecological systems. Fundamental concepts within the discipline are those of the landscape 'patch', 'edge' and 'corridor' (see Chapter 3 in this volume), and the dynamic nature of ecological patterns. In order to contrast some of the ideas in landscape ecology with what has gone before, we will review each of them here. From the previous discussion, it can be seen that much of the work often presented under the umbrella of landscape ecology (and reviewed below), does not stray very far from the conceptual focus of the last 30 years. Those studies which represent a shift in focus sufficient to begin to define landscape ecology as a separate discipline, have been grouped in the sections on 'The dynamic use of landscape' and 'Broadening the scale' (see below).

Patch size

Discussion of the effect of patch area on community structure (itself an elaboration of one of the central themes of the old paradigm) has been extensive (MacArthur and Wilson, 1967; Forman *et al.*, 1976; Whitcombe *et al.*, 1977; Connor and McCoy, 1979; Martin, 1980; Burgess and Sharpe, 1981; Butcher *et al.*, 1981; Ambuel and Temple, 1983; Lovejoy *et al.*, 1983, 1986; Woolhouse, 1983; Blake and Karr, 1984; Helle, 1984; Lynch and Whigham, 1984; Opdam *et al.*, 1985; Dobkin and Wilcox, 1986; Rosenberg and Raphael, 1986; Haila *et al.*, 1987; van Dorp and Opdam, 1987).

The message from such studies is complex, but the implication is that larger patches generally hold a greater number of species than smaller patches. Galli *et al.* (1976), for example, found 17 bird species to be present in New Jersey forest islands regardless of patch area, and 18 species to be area dependent. With increasing patch area, members of this latter group were seen to appear as their minimum habitat size requirements were met. Various mechanisms may operate to impose minimum-area requirements on bird species: minimum size for foraging needs (red bellied woodpecker and red shouldered hawk); a minimum breeding territory size or the presence of a 'sufficiently' isolated forest interior for ground nesting birds (black and white warbler, and ovenbird); and the probability of finding standing dead trees would increase with area, and facilitate the presence of hole nesters (hairy wood-pecker, black-capped chickadee and white-breasted warbler).

Species–area effects on 'habitat islands' are not always simple. Immediately after clearance of surrounding habitat, remnant patches have been found to contain more species than less fragmented stands presumably due to concentration of species from the disturbed habitat into the remnant (Lovejoy *et al.*, 1983, 1986; Rosenberg and Raphael, 1986; Wiens, 1989). More commonly, exceptions to the simple species–area relationship have been interpreted in terms of the effect of patch edge. Smaller patches have greater edge/interior ratios, and increases in species diversity after habitat fragmentation may be due to colonization of patches by edge adapted species (Lovejoy *et al.*, 1983, 1986; Gotfryd and Hansell, 1986; Temple, 1986). Fragmentation also increases patch isolation, and this has been shown to affect patch occupancy (Lynch and Whigham, 1984; Howe, 1984; Opdam *et al.*, 1985; van Dorp and Opdam, 1987; Askins *et al.*, 1987). Bird species also respond to fragmentation in different ways. Area, edge/interior ratio, isolation of patches, or all (or none) of these may be important. As edge/interior ratio and isolation are affected by area changes, concentration on patch area as a regulator of community structure, to the exclusion of other mechanisms, is unrealistic (Haila *et al.*, 1987; Wiens, 1989).

Edge

Apart from being invoked as a qualification to species–area relationships, patch edges (or ecotones) have received much attention as bird habitats which exert important controls on population and community structure, and individual success (Lay, 1938; Johnston, 1947; Galli *et al.*, 1976; Gates and Gysel, 1978; Strelke and Dickson, 1980; Gates and Mosher, 1981; Morgan and Gates, 1982; Hansson, 1983; Lovejoy *et al.*, 1983, 1986; Szaro and Jakle, 1985; Gotfryd and Hansell, 1986; Raphael, 1986; Haila *et al.*, 1987; Yahner and Scott, 1988) (see also Wiens *et al.* (1985) for a discussion of 'boundary dynamics').

Ecotones not only provide the juxtaposition of two distinct habitat types (both of which may be necessary for the activities of an edge adapted species), they are also subjected to physical conditions not present in the patch interior (Forman and Godron, 1986). Forest/clearcut ecotones in Sweden (Hansson, 1983), exert an impor-tant control on the spatial distribution of birds. The highest density of birds—predominantly tree gleaners—is seen in the first 50 m of the forest edge. 'Weakening' of the edge trees by sudden exposure to sun and rain, and a consequent increase in vulnerability to insect attacks is the probable reason for the high edge densities of birds (Hansson, 1983). Clearcut species (not generally adapted to tree gleaning) tend to avoid the edge habitat and become concentrated towards the centre of the cleared area, resulting in a high gradient of diminishing bird density across the habitat discontinuity from forest to clearcut.

Use of ecotones may also influence aspects of the breeding biology of birds. Physical conditions differ at ecotones (above), but the influence of edge on prevail-ing biotic conditions may also be of importance. For example, edges may act as channels for the movement of nest predators and brood parasites which would not normally operate in the patch interior. Gates and Gysel (1978), studying forest/field ecotones in Michigan, found that most nests were concentrated in the first 10 m of the habitat discontinuity. Increased fledgling success was strongly correlated with increasing distance from the discontinuity. Nest predation dropped from 40–50% near the edge to 5–10% away from it. Cowbird nest parasitism dropped from 15–25% to 0–5% from edge to interior. Similarly, Yahner and Scott (1988) show that predation of eggs in forested patches increased with fragmentation due to the subsequent concentration of edge adapted Corvids (crows and blue jays). The selection pressure imposed by this particular edge 'effect', (all other things equal), may have been part of the reason for the evolution of interior specialists.

Corridors

Corridors in the landscape, including hedges, shelterbelts, roads and powerlines, have also received the attention of ecologists (McAtee, 1945; Harmon, 1948; Yapp, 1973; Pollard *et al.*, 1974; Wegner and Merriam, 1979; Martin, 1980; Van der Zande *et al.*, 1980; Yoakum *et al.*, 1980; Yahner, 1981, 1982, 1983; Bongiorno, 1982; Kroodsma, 1982; Arnold, 1983; Baudry, 1984).

The construction of a corridor in the landscape may cause restructuring of the bird community utilizing the mosaic. The corridor adds a habitat type to the mosaic and creates ecotones. Kroodsma (1982) studied the avifauna of power-line corridors cut through forest in Tenessee. In this region, species observed in the forest edge but rarely in the corridor would '. . . probably have been less abundant or may have been absent without the corridor'. Similarly, species utilizing the corridor directly (yellow-breasted chat, white-eyed vireo, prairie warbler, yellow throat, towhee, indigo bunting, field sparrow, and goldfinch) would probably not have been present but for the corridor.

Increases in habitat diversity, by the addition of anthropogenic landscape elements may enhance species diversity. Arnold (1983) studied the effects of ditch, hedgerow and wooded corridors on bird numbers on agricultural land in East Anglia. Average numbers of species per site increased from 5 on arable land to 7.5 if a ditch was present, 12 with the presence of a short hedge, 17 with tall hedges and 19 with the presence of a narrow strip of woodland.

Far from being beneficial to avifauna, the introduction of corridors may have deleterious effects which may be felt over large distances. Van der Zande *et al.* (1980) found that road corridors in Holland, caused a significant decrease in the density of breeding godwit and lapwing up to a distance of 2 km from the road. The intensity of disturbance is also important. The 'disturbance distance' increased with traffic volume from 480 m with 50 cars per day to 2 km for a 54 000 car-per-day highway. Such information may be of value in future routing of roads through, or adjacent to, ornithologically important habitats.

The dynamic use of landscape

One obvious benefit of corridors is that they facilitate the movement of individuals between patches, especially if the patches and corridor are of similar composition (Wegner and Merriam, 1979; Yahner, 1982; Johnson and Adkisson, 1985; Forman and Godron, 1986; Noss, 1987; Simberloff and Cox, 1987). Conversely, habitat discontinuities, whether linear or otherwise may act as dispersal barriers to some species (Diamond, 1973; Willis, 1974; Terborgh, 1975; Wegner and Merriam, 1979;

Rolstad, 1988). Wegner and Merriam (1979), studying woods in an agricultural landscape, found that birds seldom flew between woods over open fields. Woodland species tended to move between wood and fencerow more than between any other patch types. Wood nesting birds used fencerows for foraging into adjacent fields more often than forest edge. Johnson and Adkisson (1985) show how dispersal of beech nuts between woodland patches is facilitated by use of interconnecting fencerows.

The use of patches by birds cannot realistically be considered without recognition of the effect of adjacent landscape elements. Species may use or be dependent on more than one patch type, and the pattern of patch usage may not be constant over time. Szaro and Jakle (1985) document the importance of 'adjacent' habitats to riparian bird communities in Central Arizona. Riparian birds contributed 25–30% of the birds in adjacent desert washes, and 7–15% in adjacent desert upland. This interpatch traffic was largely one way, however: desert species comprised only 1–1.5% of birds in riparian habitats, and these were largely restricted to riparian edge.

The use of landscape by bird species is dynamic. The presence or otherwise of a species in a particular patch may depend as much on the nature of adjacent patches as the proximal characteristics of the patch itself. Edges affect the density and diversity of bird communities. The relative amount of edge habitat afforded by a particular patch is dependent both on patch size and patch shape (Forman and Godron, 1986). Thus linear features in the landscape (such as power-line corridors), while increasing the representation of edge adapted bird species may be detrimental to the success of rarer interior species. Fragmentation of patches leads to changes in physical and biotic flows through and along patch boundaries. This in turn may control population and community characteristics and individual success.

Broadening the scale

The majority of the studies mentioned above are limited in geographical extent. Indeed, it might be argued that to connect some of these studies to the landscape concept as some authors do is simply to state that the study of organisms in relation to any 'object' (e.g. patch, corridor, etc.) which happens to be present in a landscape, represents a landscape ecological study. Many of the studies quoted as examples of the 'landscape ecological approach' have little to do with the 'landscape concept' as quoted by some writers in the field.

There is a need to take some of the principles of landscape ecology elucidated at fine spatial scales, and make predictions as to their ramifications at the scale of whole landscapes. The 'landscape scale' is simply that at which one considers the patterns and interaction between the various mosaic elements of patch, edge and corridor. Broadening the perspective to such a scale requires consideration of the number, type and spatial arrangement of landscape elements, and the effect of this environmental heterogeneity (in time and space) on plant and animal communities.

The concept of spatial heterogeneity and its consequences has received wide attention in the literature (e.g. Chesson, 1981, 1985; Nisbet and Guerney, 1982; Lomnicki, 1980; Stenseth 1980; Hassell, 1980; ten Houte de Lange, 1984; Freemark and Merriam, 1986; Remillard et al., 1987; Wiens, 1972). Forman and Godron (1986: 470–473), Wiens (1985; 1989: Chap. 5) review the effect of spatial heterogeneity with particular reference to vertebrates in the landscape. Without wishing to undermine the value of such work, it is interesting to note how limited these reviews are, a feature which mostly reflects the inadequacy of our knowledge about processes at the landscape scale at the present time.

Major barriers to progress in landscape ecology are problems of data collection and analysis when we move to the landscape or regional scale. As the reviews of Johnston (1990) and Johnson (1990) emphasize, however, we are perhaps fortunate that the recent emphasis on larger study areas over longer time scales has coincided with the development of geographical information systems (GIS). Coupled with the increasing power of remote-sensing systems to provide ecological information at regional scales (Wickland, 1989; Haines-Young, in press), it seems likely that spatial information systems will become an indispensable tool for the landscape ecologist.

We have seen important advances in the analysis of landscape structure using spatial information systems. Turner (1990), for example, reviews techniques for the analysis of disturbance across a landscape and its influence on landscape structure. SPAN, a program developed to quantify landscape pattern and change was used by Turner and Ruscher (1988) to determine how landscape patterns in Georgia (south-east USA) had changed in the last 50 years. They showed that the landscape had become less fragmented and more connected over time. The latter was reflected in the general decrease in edges within the area and the increase in fractal dimension, contagion and dominance in the area. The nature of these changes also varied by vegetation type and physiographical region. The forest communities, for example, increased in area and became more connected, reflecting successional change. The landscapes of the coastal plain exhibited much larger patches than either the piedmont or the mountain regions. Turner (1990) goes on to consider how such data can be used to consider the spatial spread of disturbance across a landscape, and the effect of scale on ecological processes.

Other routines for the analysis of spatial structure in landscape mosaics have been variously described by Dufourmont *et al.* (1991) and Gulink *et al.* (Chapter 10 in this volume). There is little doubt that through the use of spatial information systems, new levels of sophistication can be achieved using their analytical power. An analysis of the growing body of literature shows that GIS have now been widely employed both to model wildlife distributions (Miller *et al.*, 1989; Johnson, 1990; Walker, 1990) (see also Chapter 16 in this volume) and the structural and functional properties of ecosystems (Cook *et al.*, 1989; Kesner and Meentemeyer, 1989; Burke *et al.*, 1990; Pastor and Broshcart, 1990; Davis and Goetz, 1990). Two recent studies (Dufourmont *et al.*, 1991; Downey *et al.*, 1990) which consider issues of avian distribution are particularly interesting (see also Chapter 19 in this volume).

Dufourmont and coworkers describe how analysis of the structural patterns can assist in understanding the ecology of the barn owl (*Tyto alba*), a raptor species which is threatened in many areas of Europe by the effects of landscape change. The work was undertaken in eastern Brabant in Flanders, Belgium. The breeding success of this species at 42 nesting sites in a study area of 1400 km² was considered in relation to the structure of the landscape mosaic in a quadrat 6.25 km² around the nest site. Structural parameters such as extent of open space, patch shape complexity, size frequency distribution and edge length of woody patches were measured using data derived from classified SPOT imagery. Significant differences were found between landscapes where the barn owl breeds successfully and those where it does not. Favourable landscape characteristics included long length of view and the scarcity of roads in the neighbourhood.

Downey *et al.* (1990) report on the use of GIS and remote sensing in analysing the distribution of black grouse (*Tetrao tetrix*) in the Zdarske Vrchy region of

Czechoslovakia. This has allowed the identification of suitable areas outside the present range of the bird for possible reintroduction. Favourable factors for this bird identified by image analysis and various GIS techniques include closeness to forest edge, suitable topography, high landscape complexity as revealed by image filtering to reveal ecotones, close proximity to water and isolation from human habitation.

The studies of barn owl and black grouse are of interest because they illustrate just how the analytical capabilities of GIS can be used to explore ecological patterns in ways that would be difficult by other means. As landscape ecologists we now have a set of tools which begin to match the scale of questions we want to ask about ecological processes. In broadening the scale of investigation in landscape ecology to examine more novel and perhaps more complex concepts such as the 'ecosystem trophic modules' of Cousins (see Chapter 6 in this volume) or the 'landscape response units' described by Perez-Trejo (see Chapter 7 in this volume), the facilities which GIS offer for data storage and data integration seem even more vital.

Conclusions

There is little doubt that access to spatial information systems will be a major stimulus to landscape ecology. Perhaps the major problem which now faces us is to match these technical advances with the development of new conceptual models which describe the operation of ecological systems at the landscape scale. In developing such concepts it is unlikely that notions of equilibrium will be entirely abandoned, for at the level of the individual patch some assumption about the adjustment between pattern and process may be relevant. Ideas of equilibrium, however, will have to be looked at in the context of other processes. Assumptions that population structure can be understood in terms of a single or even several factors across its range may have to be modified. At the landscape scale, questions about the way in which organisms and landscape interact must, for example, address such issues as:

1. How the local subpopulations relate to each other and to the population as a whole (i.e. the metapopulation) across a heterogeneous landscape. Does the population behave as a single unit or a set of semi-independent units?
2. How the effect of landscape pattern on a species varies across its range. Do species occupy distinct spatial niches at the centre and edge of their ranges?
3. How populations split into subpopulations respond to environmental change through colonization and local extinctions. At what speed do the subpopulations adjust and how does this relate to the distribution of the species at the regional scale?
4. How the spatially coincident subpopulations of different species relate to each other. Do distinct communities exist, in the sense that interactions occur, or do they behave as essentially independent units?
5. How the status of the subpopulations relate to the spatial and temporal patterns of stress and disturbance across a landscape. In particular, what role does human impact have on these patterns and how is it changing?
6. The extent to which the distribution of the metapopulation is controlled by patch structure, and the extent to which it is controlled by internal processes within the population.

Turner (1990) has suggested a number of hypotheses which relate to the effects of scale on landscape pattern and process and argued that they may be tested by using GIS technology. While these ideas are interesting in their own right, the relationship between such scale effects and specific ecological processes also needs to be elucidated. Thus, in addition to the sorts of question posed above, we also need to consider issues such as:

1. What spatial scales are appropriate for the analysis of particular ecological patterns in species response? How large does a patch have to be, and how distinct from the landscape matrix before it can support a local population?
2. How do processes operating at one scale relate to patterns and processes at another? For example, while the pattern of disturbance may be spatially random, the patterns of recolonization may result in distributional shifts if these are occurring in a changing environment.

Although these and other questions posed above do not necessarily require access to spatial information systems for us to tackle them, they are made more tractable by access to such technology. For these issues will require the landscape ecologist to handle substantial volumes of data and to undertake complex analysis if they are to be attempted. Moreover, access to spatial information systems will be essential if we are to model these complex situations and test outcomes against real world data.

Whether landscape ecology as it is viewed in many parts of North America and Europe, does represent a distinct break with the mainstream ecological ideas which have dominated ecology up to the early 1980s is perhaps a matter of personal opinion. What is more certain is that the framework which the discipline represents contains a much richer view of the operation of ecological systems that has gone before. There is greater emphasis both on the effects of history and spatial structure on such systems.

As the complexity of our theories and models grows, more sophisticated ways of representing ideas and presenting descriptions of ecological systems need to be developed. In this chapter we have considered the role of spatial information systems. It would seem that access to the new technology might provide both a language and tradition in which we can better articulate the concerns of the landscape ecologist. Through its wider use we may gain deeper insight into the spatial structure and dynamics of ecological systems.

References

Ambuel, B. and Temple, S. A., 1983, Area dependent changes in the bird communities and vegetation of southern Wisconsin forests, *Ecology*, **64**, 1057–1068.

Arnold, G. W., 1983, The influence of ditch and hedgerow structure, length of hedgerows and area of woodland and garden on bird numbers on farmland, *Journal of Applied Ecology*, **20**, 731–750.

Askins, R. A., Philbrick, M. J. and Sugeno, D. S., 1987, Relationship between the regional abundance of forest and the composition of forest bird communities, *Biological Conservation*, **39**, 129–152.

Baudry, J., 1984, Effects of landscape structure on biological communities: the case of hedgerow network landscapes, in Brandt, J. and Agger, P. (Eds), *Proc. First*

International Seminar on Landscape Ecological Research and Planning, Vol. 1, pp. 55–66, Roskilde, Denmark.

Blake, J. G. and Karr, J. R., 1984, Species composition of bird communities, and the conservation benefit of small versus large forests, *Biological Conservation*, **30**, 173–188.

Bongiorno, S. F., 1982, Land use and summer bird populations in northwestern Galicia, Spain, *The Ibis*, **124**, 1–12.

Burgess, R. L. and Sharpe, D. M. (Eds), 1981, *Forest Island Dynamics in Man Dominated Landscapes*, New York: Springer-Verlag.

Burke, I. C., Schimel, D. S., Youker, C. M., Parton, W. J., Joyce, L. A. and Lauenroth, W. K., 1990, Regional modelling of grassland biogeochemistry using geographic information systems, *Landscape Ecology*, **4**, 44–54.

Butcher, G. S., Niering, W. A., Barry, W. J. and Goodwin, R. H., 1981, Equilibrium biogeography and the size of nature preserves: an avian case study, *Oecologia (Berlin)*, **49**, 29–37.

Chesson, P. L., 1981, Models for spatially distributed populations: the effect of within patch variability, *Theoretical Population Biology*, **19**, 288–325.

Chesson, P. L., 1985, Coexistence of competitors in spatially and temporally varying environments: a look at the combined effects of different sorts of variability, *Theoretical Population Biology*, **28**, 263–287.

Cody, M. L., 1981, Habitat selection in birds: the roles of vegetation structure, competitors, and productivity, *BioScience*, **31**, 107–113.

Colwell, R. K., 1984, What's new? Community ecology discovers biology, in Price, P. W., Slobodchikoff, C. N. and Gaud, W. S. (Eds), *A New Ecology: Novel Approaches to Interactive Systems*, pp. 387–396, New York: Wiley.

Colwell, R. K. and Winkler, D. W., 1984, A null model for null models in biogeography, in Strong, D. R., Simberloff, D., Abele, G. and Thistle, A. B. (Eds), *Ecological Communities, Conceptual Issues and the Evidence*, pp. 344–359. Princeton, NJ: Princeton University Press.

Connor, E. F. and McCoy, E. D., 1979, The statistics and biology of the species area relationship, *American Naturalist*, **113**, 791–833.

Cook, E. A., Iverson, L. R. and Graham, R. L., 1989, Estimating forest productivity with Thematic mapper and biogeographical data, *Remote Sensing of Environment*, **28**, 131–141.

Davis, M. B., 1986, Climatic instability, time lags and community disequilibrium, in Diamond, J. and Case, T. J. (Eds), *Community Ecology*, pp. 269–284, New York: Harper and Row.

Davis, F. W. and Goetz, S., 1990, Modelling vegetation patterns using spatial digital terrain data, *Landscape Ecology*, **4**, 69–80.

Diamond, J. M., 1973, Distributional ecology of New Guinea birds, *Science, New York*, **179**, 759–769.

Diamond, J. M., 1978, Niche shifts and the rediscovery of interspecific competition, *American Scientist*, **66**, 322–331.

Dobkin, D. S. and Wilcox, B. A., 1986, Analysis of natural forest fragments: riparian birds in the Toiyabe Mountains, Nevada, in Verner, J., Morrison, M. L. and Ralph, C. J. (Eds), *Wildlife 2000. Modelling Habitat Relationships of Terrestrial Vertebrates*, pp. 293–299, Madison, WI: University of Wisconsin Press.

Downey, I., Heywood, I., Kless, P., Pauknerova, E. and Petch, J., 1990, GIS for landscape management, Zdarske Vrchy, Czechoslovakia, *Conference Proceedings: GIS for the 1990s*, Ottawa, Canada, pp. 1528–1538.

Dufourmont, H., Andries, A., Gulink, H. and Wouters, P., 1991, A structural landscape analysis based on SPOT imagery, in *Proc. European Seminar on Practical Landscape Ecology*, International Association of Landscape Ecology, Roskilde, 2–4 May 1991.

Forman, R. T. T. and Godron, M., 1986, *Landscape Ecology*, New York: Wiley.

Forman, R. T. T., Galli, A. E. and Leck, C. F., 1976, Forest and avian diversity in New Jersey woodlots with some land use implications, *Oecologia (Berlin)*, **26**, 1–8.

Freemark, K. E. and Merriam, H. G., 1986, Importance of area and habitat heterogeneity to bird assemblages in temperate forest fragments, *Biological Conservation*, **36**, 115–141.

Galli, A. E., Leck, C. F. and Forman, R. T. T., 1976, Avian distribution patterns in forest islands of different sizes in central New Jersey, *The Auk*, **93**, 356–364.

Gates, J. E. and Gysel, L. W., 1978, Avian nest dispersion and fledgling success in field-forest ecotones, *Ecology*, **59**, 871–883.

Gates, J. E. and Mosher, J. A., 1980, A functional approach to estimating habitat edge width for birds, *American Midland Naturalist*, **105**, 189–192.

Gilbert, F. S., 1980, The equilibrium theory of island biogeography: fact or fiction?, *Journal of Biogeography*, **7**, 209–235.

Gotfryd, A. and Hansell, R. I. C., 1986, Prediction of bird community metrics in urban woodlots, in Verner, J., Morrison, M. L. and Ralph, C. J. (Eds), *Wildlife 2000. Modelling Habitat Relationships of Terrestrial Vertebrates*, pp. 321–326, Madison, WI: University of Wisconsin Press.

Haila, Y., Hanski, I. K. and Raivo, S., 1987, Breeding bird distributions in fragmented coniferous taiga in southern Finland, *Ornis Fennica*, **64**, 90–106.

Haines-Young, R. H., in press, Remote sensing, GIS and the assessment of environmental change, in Roberts, N. (Ed.), *Global Environmental Change*, Oxford: Blackwell.

Hansson, L., 1983, Bird numbers across edges between mature conifer forest and clearcuts in central Sweden, *Ornis Scandinavica*, **14**, 97–103.

Harmon, W. H., 1948, Hedgerows, *American Forester*, **54**, 448–450.

Hassell, M. P., 1980, Some consequences of habitat heterogeneity on population dynamics, *Oikos*, **35**, 150–160.

Hayes, T. D., Riskind, D. H. and Pace III, W. L., 1987, Patch within patch restoration of man-modified landscapes within Texas state parks, in Goigel-Turner, M. (Ed.), *Landscape heterogeneity and disturbance*, pp. 173–198, New York: Springer-Verlag.

Helle, P., 1984, Effects of habitat area on breeding bird communities in northeastern Finland, *Annalles Zoologici Fennici*, **21**, 421–425.

Howe, R. W., 1984, Local dynamics of bird assemblages in small forest habitat islands in Australia and North America, *Ecology*, **56**, 1585–1601.

Johnson, L. B., 1990, Analyzing spatial and temporal phenomena using geographical information systems—a review of ecological applications, *Landscape Ecology*, **4**, 31–44.

Johnson, W. C. and Adkisson, 1985, Dispersal of beech nuts by blue jays in fragmented landscapes, *The American Midland Naturalist*, **113**, 319–324.

Johnston, V. R., 1947, Breeding birds at the forest edge in Illinois, *Condor*, **49**, 45–53.

Kesner, B. T. and Meentemeyer, V., 1989, A regional analysis of total nitrogen in an agricultural landscape, *Landscape Ecology*, **2**, 151–163.

Kingsland, S. E., 1985, *Modelling Nature. Episodes in the History of Population Ecology*, Chicago, IL: Chicago University Press.

Kroodsma, R. L., 1982, Bird community ecology on power-line corridors in east Tennessee, *Biological Conservation*, **23**, 79–94.

Kuhn, T. S., 1970, *The Structure of Scientific Revolutions*, 2nd Edn, Chicago, IL: University of Chicago Press.

Lay, D. W., 1938, How valuable are woodland clearings to birdlife?, *Wilson Bulletin*, **50**, 254–256.

Lomnicki, A., 1980, Regulation of population density due to individual differences and patchy environment, *Oikos*, **35**, 185–193.

Lovejoy, T. E., Bierregaard, R. O., Rankin, J. M. and Schubert, H. O. R., 1983, Ecological dynamics of tropical forest fragments, in Sutton, S. L., Whitmore, T. C. and Chadwick, A. C. (Eds), *Tropical Rainforest: Ecology and Management*, pp. 377–384, Oxford: Blackwell Scientific Publishers.

Lovejoy, T. E., Bierregaard, R. O., Rylands, A. B., Malcolm, J. R., Quintela, C. E., Harper, L. H., Brown Jr, K. S., Powell, A. H., Powell, G. V. N., Schubart, H. O. R. and Hays, M. B., 1986, Edge and other effects of isolation on Amazon forest fragments, in Soule, M. E. (Ed.), *Conservation Biology: The Science of Scarcity and Diversity*, pp. 257–285, Sunderland, MA: Sinauer Associates.

Lynch, J. F. and Whigham, D. F., 1984, Effects of forest fragmentation on breeding bird communities in Maryland, U.S.A., *Biological Conservation*, **28**, 287–324.

MacArthur, R. H., 1957, On the relative abundance of bird species, *Proceedings of the National Academy of Sciences USA*, **43**, 293–295.

MacArthur, R. H., 1958, Population ecology of some warblers in northeastern coniferous forests, *Ecology*, **39**, 599–619.

MacArthur, R. H., 1965, Patterns of species diversity, *Biological Reviews*, **40**, 510–533.

MacArthur, R. H., 1970, Species packing and competitive equilibrium for many species, *Theoretical Population Biology*, **1**, 1–11.

MacArthur, R. H., 1972, *Geographical Ecology*, New York: Harper and Row.

MacArthur, R. H. and Levins, 1967, The limiting similarity, convergence and divergence of coexisting species, *The American Naturalist*, **101**, 377–385.

MacArthur, R. H. and MacArthur, 1961, On bird species diversity, *Ecology*, **42**, 594–598.

MacArthur, R. H. and Wilson, E. O., 1967, *The Theory of Island Biogeography*. Princeton, NJ: Princeton University Press.

Martin, T. E., 1980, Diversity and abundance of spring migratory birds using habitat islands on the Great Plains, *Condor*, **82**, 430–439.

McAtee, W. L., 1945, *The Ring Necked Pheasant and its Management in North America*, Washington, DC: American Wildlife Institute.

Miller, R. I., Stuart, S. N. and Howell, K. M., 1989, A methodology for analyzing rare species distribution patterns utilizing GIS technology: the rare birds of Tasmania, *Landscape Ecology*, **2**, 173–189.

Morgan, K. A. and Gates, J. E., 1982, Bird population patterns in forest edge and strip vegetation at Remington farms, Maryland, *Journal of Wildlife Management*, **46**, 933–944.

Nisbet, R. M. and Gurney, W. S. C., 1982, *Modelling Fluctuating Populations*, New York: Wiley.

Noss, R. F., 1987, A regional landscape approach to maintain diversity, *BioScience*, **33**, 700–705.

Opdam, P. F. M., Rijsdijk, G. and Hustings, F., 1985, Bird communities in small woods in an agricultural landscape: effects of area and isolation, *Biological Conservation*, **34**, 333–352.

Pastor, J. and Broschart, M., 1990, The spatial pattern of northern conifer hardwood landscape, *Landscape Ecology*, **4**, 55–68.

Pollard, E., Hooper, M. D. and Moore, N. W., 1974, *Hedges*, London: Collins.

Price, P. W., 1984, Alternative paradigms in community ecology, in Price, P. W., Slobodchikoff, C. N. and Gaud, W. S. (Eds), *A New Ecology: Novel Approaches to Interactive Systems*, pp. 353–386, New York: Wiley.

Price, P. W., Slobodchikoff, C. N. and Gaud, W. S., 1984, Introduction: is there a new ecology?, in Price, P. W., Slobodchikoff, C. N. and Gaud, W. S. (Eds), *A New Ecology: Novel Approaches to Interactive Systems*, pp. 1–14, New York: Wiley.

Remillard, M. M., Gruendling, G. K. and Bogucki, D. J., 1987, Disturbance by beaver (*Castor canadensis* Kuhl) and increased landscape heterogeneity, in Goigel-Turner, M. (Ed.), *Landscape Heterogeneity and Disturbance*, Berlin: Springer-Verlag.

Rolstad, J., 1988, Autumn habitat of Capercaillie in south east Norway, *Journal of Wildlife Management*, **52**, 747–753.

Rosenberg, K. V. and Raphael, M. G., 1986, Effects of forest fragmentation on vertebrates in Douglas fir forests, in Verner, J., Morrison, M. L. and Ralph, C. J. (Eds), *Wildlife 2000. Modelling Habitat Relationships of Terrestrial Vertebrates*, pp. 263–272, Madison, WI: University of Wisconsin Press.

Rotenberry, J. T. and Wiens, J. A., 1980a, Temporal variation in habitat structure and shrubsteppe bird dynamics, *Oecologia*, **47**, 1–9.

Rotenberry, J. T. and Wiens, J. A., 1980b, Habitat structure, patchiness, and avian communities in North American steppe vegetation: a multivariate analysis, *Ecology*, **61**, 1228–1250.

Simberloff, D. and Cox, J., 1987, Consequences and costs of conservation corridors, *Conservation Biology*, **1**, 63–71.

Stauffer, D. F. and Peterson, S. R., 1985, Ruffed and blue grouse habitat use in south east Idaho, *Journal of Wildlife Management*, **49**, 459–466.

Stenseth, N. C., 1980, Spatial heterogeneity and population stability: some evolutionary consequences, *Oikos*, **35**, 165–184.

Strelke, W. K. and Dickson, J. G., 1980, Effect of forest clearcut edge on breeding birds in east Texas, *Journal of Wildlife Management*, **44**, 559–567.

Szaro, R. C. and Jakle, M. D., 1985, Avian use of a desert Riparian island and its adjacent scrub habitat, *The Condor*, **87**, 511–519.

Temple, S. A., 1986, Predicting impacts of habitat fragmentation on forest birds: a comparison of two models, in Verner, J., Morrison, M. L. and Ralph, C. J. (Eds), *Wildlife 2000. Modelling Habitat Relationships of Terrestrial Vertebrates*, pp. 301–304, Madison, WI: University of Wisconsin Press.

ten Houte de Lange, S. M., 1984, Effects of landscape structure on animal population and distribution, in Brandt, J. and Agger, P. (Eds), *Proc. First International Seminar on Landsacpe Ecological Research and Planning*, Vol. 1, pp. 19–31, Roskilde, Denmark.

Terborgh, J., 1975, Faunal equilibria and the design of wildlife preserves, in Golley, F. B. and Medina, E. (Eds), *Tropical Ecological Systems: Trends in Terrestrial and Aquatic Research (Ecological Studies II)*, pp. 369–380, New York: Springer-Verlag.

Turner, M. G., 1990, Spatial and temporal analysis of landscape patterns, *Landscape Ecology*, **4**, 21–30.

Turner, M. G. and Ruscher, 1988, Changes in landscape patterns in Georgia, USA, *Landscape Ecology*, **1**, 241–251.

van Dorp, D. and Opdam, P. F. M., 1987, Effects of patch size, isolation and regional abundance on forest bird communities, *Landscape Ecology*, **1**, 59–73.

van der Zande, A. N., ter Keurs, W. J., and van der Weijden, W. J., 1980, The impact of roads on the densities of four bird species in an open field habitat: evidence of a long distance effect, *Biological Conservation*, **18**, 299–321.

Walker, P. A., 1990, Modelling wildlife distributions using a geographic information system: kangaroos in relation to climate, *Journal of Biogeography*, **17**, 297–289.

Wegner, J. F., and Merriam, G., 1979, Movements by birds and small mammals between a wood and adjoining farmland habitats, *Journal of Applied Ecology*, **16**, 349–358.

Whitcombe, B. L., Whitcombe, R. F. and Bystrak, D., 1977, Island biogeography and 'habitat islands' of eastern forests. III Long term turnover and effects of selective logging on the avifauna of forest fragments, *American Birds*, **31**, 17–23.

Wickland, D. E., 1989, Future directions for remote sensing in terrestrial ecological research, in Asrar, G. (Ed.), *Theory and Applications of Optical Remote Sensing*, pp. 691–724, New York: Wiley.

Wiens, J. A., 1972, Population responses to patchy environments, *Annual Review of Ecology and Systematics*, **7**, 81–120.

Wiens, J. A., 1984, On understanding a non-equilibrium world: myth and reality in community patterns and processes, in Strong, D. R., Simberloff, D., Abele, L. G. and Thistle, A. B. (eds), *Ecological Communities, Conceptual Issues and the Evidence*, pp. 439–509, Princeton, NJ: Princeton University Press.

Wiens, J. A., 1985, Vertebrate responses to environmental patchiness in arid and semi-arid ecosystems, in Pickett, S. T. A. and White, P. S. (Eds), *Natural Disturbance and Patch Dynamics*, pp. 169–193, New York: Academic Press.

Wiens, J. A., 1986, Spatial scale and temporal variation in studies of shrub-steppe birds, in Diamond, J. and Case, T. J. (Eds), *Community Ecology*, New York: Harper and Row.

Wiens, J. A., 1987, Habitat occupancy patterns of North American shrub-steppe birds: the effects of spatial scale, *Oikos*, **48**, 132–147.

Wiens, J. A., 1989, *The Ecology of Bird Communities*, Vols 1 and 2, Cambridge: Cambridge University Press.

Wiens, J. A., Crawford, C. S. and Gosz, J. R., 1985, Boundary dynamics: a conceptual framework for studying landscape ecosystems, *Oikos*, **45**, 421–427.

Williamson, K., 1975, Birds and climatic change, *Bird Study*, **22**, 143–165.

Willis, E. O., 1974, Population and local extinctions of birds on Barro Colorado Island, Panama, *Ecological Monographs*, **44**, 153–169.

Woolhouse, M. E. J., 1983, The theory and practise of the species-area effect applied to breeding birds of British woods, *Biological Conservation*, **27**, 315–332.

Yahner, R. H., 1981, Avian winter abundance patterns in farmstead shelter belts: weather and temporal effects, *Journal of Field Ornithology*, **52**, 50–56.

Yahner, R. H., 1982, Avian use of vertical structures and plantings in farmstead shelter belts, *Journal of Wildlife Management*, **46**, 50–60.

Yahner, R. H., 1983, Seasonal dynamics, habitat relationships and management of avifauna in farmstead shelter belts, *Journal of Wildlife Management*, **47**, 85–104.

Yahner, R. H. and Scott, D. P., 1988, Effects of forest fragmentation on depredation of artificial nests, *Journal of Wildlife Management*, **52**, 158–161.

Yapp, W. B., 1973, Ecological evaluation of a linear landscape, *Biological Conservation*, **5**, 45–47.

Yoakum, J., Dasman, W. P., Sanderson, H. R., Nixon, C. M. and Crawford, H. S., 1980, Habitat improvement techniques, in *Wildlife Management Techniques Manual*, pp. 344–345, Washington, DC: Wildlife Society.

6

Hierarchy in ecology: its relevance to landscape ecology and geographic information systems

S. H. Cousins

The objective of this chapter is to unravel some of the complexity which obscures our understanding of the landscape and its ecology. If this objective is successful it will redefine what are the important requirements for data collection and so help structure GIS for landscape applications.

To say that landscape ecology is interdisciplinary is an understatement. While cross-discipline studies are generally seen as scientifically creative, they are also problematic. It is characteristic of such studies that there is a conceptual lag in taking new developments across discipline boundaries; geographers use old ecology, while ecologists use old geographical ideas, and so on. True to this approach, as an ecologist I report some very recent progress in ecological science and then combine it with some rather less fashionable geographical ideas from the 1930s. The result is, I hope, a much clearer picture of what ecosystems actually are. This allows a simplification of the human and biospheric interactions with ecosystems and thus a clearer perception of landscape ecology itself.

The three groups of processes identified above (the ecological, the human and the biospheric) are, in broad terms, the set of processes central to landep units—namely, the ecological complex, the anthropo-complex and the ambio-complex, respectively (see Chapter 5 in this volume). The potential to simplify at least the conceptual complexity of landscape ecology comes from the relatively recent use of hierarchy theory.

Hierarchy theory: a tool-kit for geographic information systems

Naveh and Lieberman (1983) provides a definition of landscape which captures both the variety of scale and the interdisciplinarity of landscape ecology. He says that landscapes are 'a part of the space on the Earth's surface, consisting of a complex of systems, formed by the activity of rock, water, air, plants, animals and man [which] forms a recognisable entity'. Hierarchy theory allows the decomposition of these 'complexes' into strongly and weakly interacting components. Indeed, one of the central tenets of hierarchy theory (Simon, 1973) is that objects at one level in the hierarchy are nearly independent of objects at levels below and above it and so are weakly connected in those directions, while connections at the same level are much stronger.

To apply hierarchy theory to landscape problems, some simple conceptual tools are needed. One tool has already been noted, that is the quasi-independence of objects at different hierarchical levels. However, to use this idea a second tool is needed; objects have to be clearly defined and indeed clearly separated from non-objects such as aggregates. Rowe (1961) shows that objects are organized as containing structurally connected parts, while aggregates occupy a common area, but have no structural organization. Thus, although a forest may appear as a solid object when viewed from a distance, Rowe contends it is an aggregate of objects (the plants), but is not an object itself. In comparison, the biological hierarchy of cell–organ–oganism–ecosystem is a hierarchy of objects where each object contains structurally related parts; thus the organ is composed of cells, the organism made up of organs and, as is shown later, the ecosystem is composed of organisms.

The choice of scale of observation is important in landscape ecology (see, for example, Chapter 8 in this volume). Questions of scale are also highly relevant to the distinction between aggregates and objects noted above. Objects do have intrinsic scale, if only within broad limits, whereas aggregates do not. Thus a forest can be of any size greater than a certain basic size, the individual objects (the trees) are of a characteristic size, given particular external environmental constraints and internal biological constraints.

As well as the awareness that a change in scale changes the number and extent of what is observed, it also affects the types of phenomena that can be observed. A significant change in scale is therefore associated with hierarchies of phenomena. This is perhaps most clearly seen in biological systems where observation using microscopes reveals subcellular organization, then cells and then, by direct observation, organs, bodies and so on, while observation from space is required to see the biosphere.

Hierarchies of scale, by which we mean scale of observation, present different issues because scale can be changed in a continuous manner. This leads to the final tool introduced here which concerns the importance of distinguishing types of hierarchy. It is important to do this to ensure that different types are not mixed. Thus a single hierarchy should not contain aggregates and integrated objects. Distinguishing different types of hierarchy allows the interpretation of what the hierarchies mean.

What is an ecosystem?

Landscape ecologists could be forgiven for thinking that textbook definitions of 'ecosystem' could be imported for use in the analysis of landscape phenomena. However, as shown below, the concept of an ecosystem is currently under re-evaluation within the science of ecology. This debate can be examined by looking at a traditional definition of 'ecology', its problems and the proposed solutions.

Perhaps the most influential definition of 'ecosystem' developed this century has been that of Lindeman (1942) given as part of the introduction to his famous

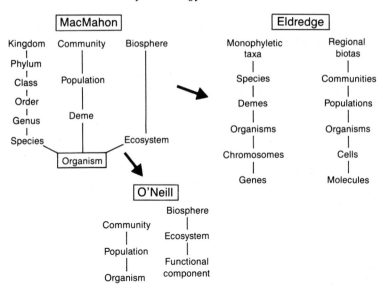

Figure 6.1. Some recent approaches to classifying the components of biological organization.

paper on trophic levels. Lindeman takes the position that:

> the ecosystem can be formally defined as the system composed of physical–chemical–biological processes active within a space time unit of any magnitude, i.e. the biotic community and its environment.

The two main features of this definition, that the ecosystem is composed of biotic and abiotic parts, and that the ecosystem has no intrinsic scale, have both become the accepted wisdom of our time. However, this view that ecosystems are, in a sense, everything and present at any scale, has led many ecologists to question the reality of the ecosystem concept. O'Neill *et al.* (1986) observe that 'the ecosystem as an independent discrete entity looks less and less tenable', while Ghiselin (1987) points to this kind of limitation at the root of ecological science, stating that 'ecologists are most unsure about the nature of their fundamental units and about what such units do'.

It is this search for basic ecological units (components) that has led many ecologists to construct hierarchies of biological phenomena in search of plausible candidates (see Figure 6.1). This is plainly a varied set of hierarchies, with each based on different types of relationship. The taxonomic hierarchies of species to kingdoms are linked by the history of evolutionary descent and are not, at each or any of the levels, functioning objects today (Grene, 1987). The hierarchies of community–population–deme and biosphere–ecosystem–organism are different in many ways (Cousins, 1988), principally because the community hierarchy is one of concepts rather than physical objects. Thus the deme (a local breeding population with high mutual gene flow), a population (a collection of demes) and a community of populations of many species all have boundaries which are subjectively chosen by the observer. The biosphere (later called the 'Earth-biosphere') and individual are, within certain limits, objectively defined functional objects which are independent of the observer and can 'do' things in the sense called for by Ghiselin (1987). Although the key concept of the ecosystem is again a subjectively determined aggregate with boundaries given by an observer, it is possible to define an ecological object which substitutes for ecosystem in a hierarchy of functional objects.

The ecosystem trophic module (the ecosystem object)

Biological systems are, like landscapes, far from thermodynamic equilibrium. In such systems, structure (or, more precisely, organization) is created by the passage of energy through the system. In ecosystems the path of energy dissipation and material flows is determined by feeding and respiration. It is as a result of photosynthesis by plants and feeding by herbivores, by carnivores and by decomposers that an ecological structure is created. This structure is the distribution of organisms in space, including the relative abundance or biomass of different types of organism. If we imagine solar radiation incident on an ecosystem, then the path the energy takes is the familiar one of being captured by the plant and then is either reradiated as heat or passes to decomposers or to herbivores and carnivores. Spatially, what occurs is that energy captured by the plant is first used by the plant to concentrate what was the uniform field of solar energy into a variety of energy states including energy-dense sources such as seeds, down to energy-poor leaf drip (Cousins, 1980). These energy sources are then further concentrated, dispersed, or respired by herbivores and detrivores such that some of the energy reaches the top predator.

Figure 6.2 shows a diagram of energy flow directions where sunlight falls evenly over a number of contiguous territories of a top predator social group, perhaps a breeding pair of foxes or a pride of lions. Little of the energy incident on the whole territory reaches the top predator, but it enters the food chain which leads to the top predator as soon as energy has been captured by the green plant. Energy which falls on one side of the territory boundary goes to one predator; energy which falls across the boundary flows towards the adjacent predator group.

It is these structures which are bounded in space by territorial behaviour, bounded in time by the initiation and termination of the territorial unit, and which are made up of all the locally interacting organisms which form the food web of the top predator, which forms the largest ecological object at any one point on the Earth. Cousins (1988, 1990) has called this object the Ecosystem trophic module (ecotrophic module or ETM; 'trophic' meaning feeding).

The ETMs of a region of the Serengeti plain in Tanzania are shown in Figure 6.3. Note that the spacing of the lion prides which define the ETMs as shown is of the order of 10 km, giving an area of approximately 100 km^2 as the size of the

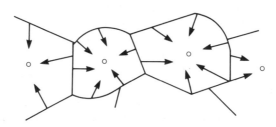

Figure 6.2. The boundaries of the ecosystem object, the ETM, formed by the overall paths of energy flow from incident solar to the social group of the top predator. (○) Centre of range of top predator social group. Arrows indicate the direction of flow of energy.

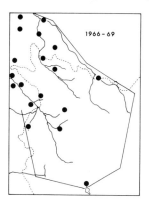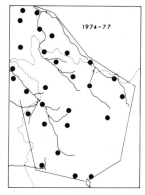

Figure 6.3. Ecotrophic modules identified by centres of prides (●) of lions on the Serengeti plains (Hanby and Bygott, 1979). (– – – –) Separates woodland in the north-west from the plains; (———) the area studied. The whole map is of an area of approximately 85 km × 60 km. (Reproduced with permission from Cousins (1990).)

ecological object. Note too that the distributions differ between the two periods shown. Hanby and Bygott (1979) suggest that this change is due to a change in the environment, namely a change in rainfall. This change in climate arises at a different hierarchical level from the behaviour of the global weather system which impacts on the ETM structure of the Serengeti.

Finally, it is important to address the question of the scale of observation in the context of the ETM. Choosing a scale of observation is always a subjective judgement based on the type of problem that is being analysed. The recognition of the ETM does not change this, as the chosen scale will include one, less than one, or more than one ETM. The size of the ETM is just one more factor to add to those governing the choice of the observation scale.

The human dimension

The human dimension is particularly important to landscape ecology. The definition of an ecosystem as a bounded ecological object, the ETM, raises the important question of where humans fit into such a structure. Are humans top predators, or do human social systems represent an entirely new level of energy flow and organization? Certainly human social groups with weapons have primitively acted as top predators and may still do so today in hunter–gatherer societies. Subsequent human groups differ from the top predator group by one very important activity, they engage in trade. Thus, whereas in the ETM the energy and materials are organized within the spatial unit circumscribed by the ETM boundary (Figure 6.2), trading humans exchange energy and materials *between* ETMs, thereby creating a new and larger entity or organization.

Geographers have long studied the spatial structures generated as a result of trade. An example of one such structure is the hierarchy of market settlements from village, local town, regional towns to cities. Christaller (see Haggett, 1965) proposed a seven-tier hierarchy of settlement from hamlet to world city in which there are an increasing range of specialist functions undertaken at each larger scale. This is the central place theory (Haggett, 1965) in which large centres of population have a much greater range of goods, services and functions than smaller centres. Large cities consume resources that are mainly produced by smaller settlements and farms (see Figure 6.4).

By undertaking trade, humans have over the last 3000 years left the ecological scale of the ETM (< 100 km^2) and are now part of a global system of world trade which we can call the 'econosphere' by analogy with the biosphere concept. Humans appear to have literally changed hierarchical level having left the ETM and created a new organization or object at the global level.

Ironically, this identification of humans with a higher level of organization than the ETM allows us to propose an objective measure of human impact. This method of environmental-impact assessment is to measure the reduction in the size of the impacted ETM compared with the preimpacted state and, ultimately, with the ETM area of the top-predator characteristic of that part of the Earth.

As one indication of the impact of human activity on ecosystems, Figure 6.5 shows the average number of species of bird found within a 20-mile radius of the centre of London and a transformation of that information showing the average species body weight of those species. For the methods used, see the Appendix. Although these data are for all bird species rather than a count of the number or size of the ETMs, they do show a net decline in body size towards the centre commensurate with a decline in energy availability to territorial birds in cities caused by limited green space and other factors such as pollution (Cousins, 1982). These maps may also be considered as a representation of an intersection of human and natural systems.

Finally, it is important to emphasize that trading or economic activities are framed within human-value systems which reinforce or override economic activities and the spatial patterns created by trade. Allen *et al.* (1984) have modelled these

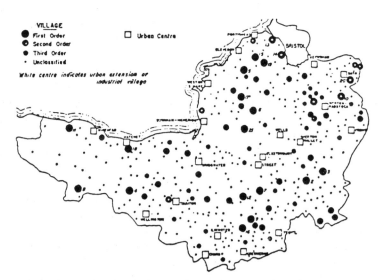

Figure 6.4. A hierarchy of settlements in Somerset, south-western England. The classification of villages is based on a continuum with breaks at 5, 10 and 20 shops. Urban centres are distinguished from villages and the city of Bristol.
(Reproduced with permission from Haggett (1965).)

processes and derived a city structure from microscale decision-making. As a further example, Allen (personal communication) has suggested that the difference in English and French aesthetic preferences for urban living, characterized as a greater preference of rich French to live in urban centres while rich English tend to prefer rural outposts, has materially affected urban structures in the two countries.

Earth biosphere

It is traditional for biologists to call the Earth, the 'biosphere'. The one word 'biosphere' encapsulates the proposition that both the surface of the Earth and its atmosphere have been radically altered by the presence of life (Rambler *et al.*, 1989). Where the term 'biosphere' is weaker is that it obscures the processes of the deeper geology, and so volcanism, for example, is left out of the equation of the biologists' atmosphere. The term 'biosphere' also obscures the recent development of the econosphere, recent that is in terms of biological and planetary history.

I shall use the term 'Earth biosphere' as a stimulus to a clearer understanding of the structure which is created by energy flows involving the Earth. We are not interested in these in detail, but only in the kind of phenomena concerned in order to find where 'landscape' fits into such a scheme.

In Figure 6.1 the hierarchy, organism–ecosystem–biosphere is replaced here by a hierarchy of organized objects, **Organism–ETM–Earth**, where the organization is achieved by flows of energy and materials. The question of interest for landscape ecology is whether there are other discrete organized objects which constitute component parts of the Earth and which are not the ETM.

Primarily, hierarchy theory allows for the creation of a new level of organization out of the interaction of a number of different parts. In the case of the body there were components of different kinds, for example organs of different types. For the Earth biosphere we may identify a series of components which together interact to form the new unit. These include, the ETM, sea and atmosphere circulation patterns, surface-water runoff, plate tectonics, and volcanism. The Earth is then part of a planetary system with gains from and loses to space.

The energy required to maintain the observed structure of the Earth comes from two principal sources, the cooling of the Earth's core and the heating up of the Earth from space by solar energy. Although we are almost exclusively interested in the structures created by solar-energy transfer, the outgassing of volcanoes adds new material to the atmosphere and can also affect the energy balance due to volcanic dust altering the Earth's albedo. To this list of the Earth's components we must also add structures created by human trade, which are treated here as contiguous city hinterlands.

With regard to the landscape without human intervention, there are two main implications of Earth-biosphere level phenomena. We need to look for structures created by energy dissipation at a scale larger than or of the same order as the ETM. We therefore see air and water movement as the structuring forces driven by temperature gradients created by differences in surface reflectance (albedo) and by evaporative cooling. Alterations in albedo, in greenhouse gas composition of the atmosphere, or other sources of climate changes, results in different rainfall levels and in different drainage amounts or patterns. On land, therefore, the landscape unit of observation is the river basin and the organizing principle is water flow from

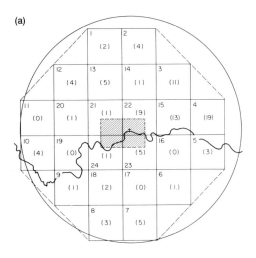

(a)

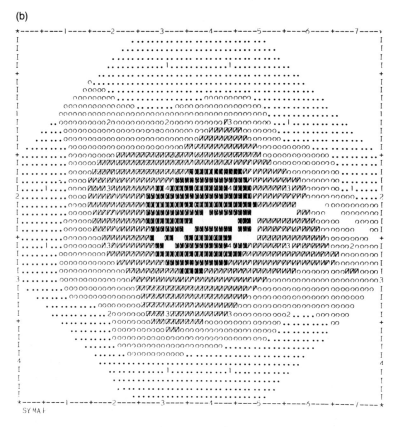

(b)

SYMAF

Figure 6.5. (a) The study zone of 24 10 km × 10 km grid squares (plus central rectangle) is shown within a radius of 32.5 km (25 miles) of St Pauls Cathedral (marked by a cross). In (b) to (d) the data are interpolated to the hatched lines. The river Thames is shown. (b) Built environment index of settlements in London. Contours 1–4 (light to dark): 3–10.9 points, 11–18.9 points, 19–26.9 points, 27–35 points. Points given for housing density; for definition of units see Cousins (1982). (c) Land bird species density/100 km². Contours 1–4 (light to dark): 43–51 species, 52–60 species, 61–68 species, 69–77 species. (d) Land bird average species weight/100 km². Contours 1–4 (light to dark): 90.5–98.5 g, 98.6–106.5 g, 106.6–114.5 g, 114.6–122.5 g.

(Reproduced with permission from Cousins (1982).)

(c)

(d)

rain falling to evaporative losses and flow of water through surface and below surface systems to the oceans. Water in this sense is a common currency which is distributed around the Earth at different rates, depending on the nature of the ground surface and the temperature regime created by the composition of the atmosphere. Water vapour is a major component of the temperature regulation of the Earth, either as low-level clouds which reduce cooling or high-level reflective clouds which increase cooling.

It may not appear at first that a great simplification of that long series of interactions identified by Zonneveld (1979) has been achieved. However, we have concluded here that three separate components have been differentiated such that the natural ecosystem is organized in ETMs, the land surface is organized in river basins, and human trade and settlement pattern is organized as cities and their hinterlands. It has been proposed that these structures together with volcanoes, atmospheric and oceanic circulation patterns, crust movements and interior structures of the planet, create a structure at a particular hierarchical level, the Earth today.

Data collection and the GIS

This chapter began with the aim of simplifying observation. What we choose to observe is, as discussed for the ETM, determined by the problems we wish to solve. We may observe part of, rather than a whole, river basin, part of the economy of a distant city and not the whole hinterland, and so on.

By looking at the landscape as an interaction of other discrete organized units we can see not only new ways of classifying observations, but also of identifying the requirements for new data and techniques for gathering it. The observation of plants by remote sensing provides an example where this new means of classification can be applied. Such remote-sensing information is conventionally seen as an ecosystem description. However, as discussed here, ecosystem units are determined by the top predator social groups and, therefore, are not visible by normal remote-sensing methods. In this context plants can be seen structures which pump water from the soil to the atmosphere. The ecosystem objects require direct observation and the data must be entered into the GIS. Similarly, if human trading activity is central to the landscape, then some form of representation of this is required in the landscape GIS if a process model of the landscape is to be created.

Perez-Trejo (see Chapter 7 in this volume) argues that process models of landscapes are essential if GISs are to be truly useful. In a given landscape, the ETM, the watershed and the trading structure lead to particular intersections of these three components at any one place and time. Perez-Trejo names regularities in this intersection as 'landscape response units'.

In the evolution of the Earth to the form in which it is found today, we have passed from a lifeless Earth to an Earth which incorporates life and is called the 'biosphere'; to an Earth which has life including trading human social groups and called here the 'econosphere'; to an Earth which has humans with values other than purely economic ones which has been called the 'noosphere' (Vernadsky, 1945). This last term serves to remind us of aesthetic values of the landscape either within the ETM, river basin or a city framework. Perhaps aesthetic values too can be placed onto the landscape GIS.

Appendix

Species Atlas data from Montier (1977) were used as the primary data input for the study. Montier's atlas records birds seen in London during the breeding season in 2 km × 2 km grid squares over the greater London area. During data capture from the atlas, the 2 km × 2 km information was aggregated into 10 km × 10 km squares. A suitably sized grid for 10 km × 10 km squares was marked on a transparency and data were taken from the Atlas by overlaying a grid on each of the single-species distribution maps and noting the presence of that species in the grid cells. This method of data capture is made more efficient by numbering the grid cells in a way which mirrors the general distribution of all species (in this case as a spiral from outside to the centre). Then data can be entered efficiently using a program which accepts information on the contiguous distribution of each species. Thus, if a species is found in all 10 km × 10 km squares except the central five, then this can be entered as 1, 20. The data are then stored in an array as a single-species row of presence (1)/absence (0) information for the grid cells.

The generation of an array of locationally independent data about the individual species allows the transformation of the presence/absence data to create an output array. In the case shown, the individual body weight of each species is used as a species variable and the output array is created by multiplying the species presence/absence array by the species body weight. Summing the number of species in each grid square gives the data shown in Figure 6.3 while summing the body weights in each grid and dividing by the number of species gives the mean species size data shown in Figure 6.4. This technique has also been applied to the distribution of breeding birds over the UK (Cousins, 1989).

References

Allen, P., Engelen, G. and Sanglier, M., 1984, Self-organising dynamic models of human systems, in Frehland, E. (Ed.), *Synergetics: From Microscopic to Macroscopic Order*, pp. 150–171, Berlin: Springer-Verlag.

Cousins, S. H., 1980, A trophic continuum derived from animal size, plant structure and a detritus cascade, *Journal of Theoretical Biology*, **82**, 607–618.

Cousins, S. H., 1982, Species size distributions of birds and snails in an urban area, in Bornkamm, R., Lee, J. A. and Seaward, M. R. D. (Eds), *Urban Ecology*, pp. 99–109, Oxford: Blackwell Scientific.

Cousins, S. H., 1988, Fundamental components in ecology and evolution, in Wolff, W. F., Soeder, C. J. and Drepper, D. R. (Eds), *Ecodynamics*, pp. 60–68, Berlin: Springer-Verlag.

Cousins, S. H., 1989, Species richness and the energy theory, *Nature*, **340**, 350–351.

Cousins, S. H., 1990, Countable ecosystems deriving from a new foodweb entity, *Oikos*, **57**, 270–275.

Eldredge, N., 1985, *Unfinished Synthesis*, Oxford: Oxford University Press.

Ghiselin, M. T., 1987, Hierarchies and their components, *Palaeobiology*, **13**, 108–111.

Grene, M., 1987, Hierarchies in biology, *American Scientist*, **75**, 504–510.

Haggett, P., 1965, *Locational Analysis in Human Geography*, London: Edward Arnold.

Hanby, J. P. and Bygott, J. D., 1979, Population changes in lions and other predators, in Sinclair, A. R. E. and Norton-Griffiths, M. (Eds), *Serengeti: Dynamics of an Ecosystem*, pp. 249–262, Chicago, IL: University of Chicago Press.

Lindeman, R. L., 1942, The trophic–dynamic aspect of ecology, *Ecology*, **23**, 399–418.

MacMahon, J. A., Phillips, D. L., Robinson, J. V. and Schimpf, D. J., 1978, Levels of biological organisation: an organism centered approach, *Bioscience*, **28**, 700–704.

Montier, D. J., 1977, *Atlas of the Breeding Birds of the London Area*, London: Batsford.

Naveh, Z. and Lieberman, A. S., 1983, *Landscape Ecology*, Berlin: Springer-Verlag.

O'Neill, R. V., DeAngelis, D. L., Wade, J. B. and Allen, T. F. H., 1986, *A Hierarchical Concept of Ecosystems*, Princeton, NJ: Princeton University Press.

Rambler, M. B., Fester, R. and Margulis, L., 1989, *Global Ecology*, New York: Academic Press.

Rowe, J. S., 1961, The level-of-integration concept in ecology, *Ecology*, **42**, 420–427.

Simon, H. A., 1973, The organisation of complex systems, in Pattee, H. H. (Ed.), *Hierarchy Theory*, pp. 1–27, New York: Braziller.

Vernadsky, W., 1945, The biosphere and noosphere, *American Scientist*, **33**, 1–12.

7

Landscape response units: process-based self-organising systems

F. Perez-Trejo

Introduction

The relationship between landscape ecology and geographic information systems (GISs) is a very interesting one to consider. The former is theoretical in nature, while the latter is a more technological development of the last decade. Landscape ecology could be seen as providing the theoretical basis for the applications developed in GIS. However, GIS can help in perceiving how spatial patterns occur and how they change over time, providing an insight into the complex interactions between physical, climatic, biological, ecological and human processes, in order to provide new theories and to expand the understanding of landscape dynamics.

The development of scientific thinking in landscape ecology in the past 20 years could be described as an amalgamation of several disciplines, such as geography, ecology, biogeography, phyto-sociology, and remote sensing. The concept of the ecology of landscapes comes from the realization that living systems (including human systems) and physical systems interact on different temporal and spatial scales to generate the landscape with all its elements and structures, and from attempting to develop a theoretical basis for that perception of the non-random structuring of natural and man-made systems.

How the landscape is perceived is central to the development of a theory that can explain what the landscape is, how it is used (or misused), and serves as a basis for thinking about what the landscape might become in the future. Considering the extent of current environmental problems such as land degradation, erosion, desertification and pollution, it is clear that GIS is an essential tool for the process of assessing and monitoring the impact of human activity and settlement patterns on spatial patterns and ecosystem dynamics, and for manipulating and displaying the information in ways that can be easily understood by those involved in studying or planning the landscape and its use.

Clearly, the dynamic processes occurring in a landscape are very complex. Ecological, climatic, physical, and socio-economic processes are linked together on different temporal and spatial scales in a complex dynamical system that has evolved into the landscape patterns that exist today. To make some theoretical sense of all this complexity, landscape ecological theory has been defined by Naveh and Lieberman (1984) as 'the study of landscape units from the smallest landscape cell to the global ecosphere landscape in their totality as ordered ecological, geographical, and cultural wholes'. Naveh and Lieberman go beyond the biological hierarchical context of the landscape on which Urban *et al.* (1987) have focused, and emphasize the important role of human ecology in understanding landscape dynamics.

This constitutes a new departure from a narrower view of landscape ecology put forward by Forman and Godron (1986), termed here 'the equilibrium view', as being a combination of physical and biological components linked by flows of energy, nutrients, water, etc., in a fixed structure of ecological populations and communities. The equilibrium view of landscape ecology perceives climatic, biological or human activity as disturbances or perturbations around this equilibrium, exogenous to the system, driving it eventually to some new equilibrium state.

The work of Prigogine and his group at the Free University of Brussels has given rise to a new scientific paradigm regarding complex systems and dissipative structures (Nicolis and Prigogine, 1977; Prigogine and Stengers, 1987). Complex-systems theory is about understanding the creative power in self-organizing systems, and it lays the theoretical foundation for understanding the processes by which matter, organisms and ecosystems structure themselves into complex structures, far from equilibrium (Prigogine *et al.*, 1977).

Even though it might be widely accepted that complex natural systems are the product of an evolutionary process, the implications of the process of evolution on the form and function of the components of landscapes are yet to be completely understood. This chapter is not concerned with debating the actual mechanisms of the evolutionary process in natural systems, but rather the way in which system-level responses can be explained by the interactions and feedback of the components in an evolutionary dynamic. Allen (1988) points out that one of the consequences of complex systems resulting from an evolutionary process is that the components of such systems are highly interrelated. The resources that one set of organisms depend upon for survival are other living organisms that are changing and adapting dynamically themselves. Adaptations or strategies respond to fluctuations or changes in each organism's environment. As each organism has as part of its environment other organisms that are changing and adapting, change and variability in strategies are prominent features of the system's dynamics. The consequence of this realization is that the dynamics of natural systems are not fluctuating around some equilibrium point, but instead are capable of what Allen *et al.* (1985) describe as qualitative changes in their structure that are self-generated by the interaction of elements within the system.

The description of landscapes

The difficulties of describing landscapes becomes most apparent in considering the challenge of describing something that is changing in its very nature. The analogy of a caterpillar that is undergoing metamorphosis can help to illustrate the problem. The difficulty resides, on the one hand, in that depending on what scale is chosen (temporal or spatial) the description will always contain a subset of the components and processes in the system. These scales are often imposed rigidly by the types of methods and tools that are used in the description, such as satellite images, aerial photos, or seasonal field studies, which might miss the temporal or spatial scales of important flows that may be relevant to the changes being studied. On the other hand, there is the problem of trying to measure and characterize the features in the system that can help to provide an indication of what the system could become under different sets of conditions. For example, a GIS could be used to determine the human population growth centres in the Mediterranean region to predict the areas that are most susceptible to pollution, using current population distributions, and existing overlays of industrial activity and sewage dumping. But the reality that will unfold in the future will be quite different. This is because population densities are affected by migration patterns, and these in turn are affected by availablility of jobs, housing and other factors. Furthermore, these could all change as trade patterns change in Europe after 1992, or due to increases in energy prices.

Part of the difficulty in characterizing the essence of the landscape entails measuring some aspect of its current state that could provide an indication of its potential to become something else that is qualitatively different from its present state. The study of the landscape must therefore include developing an understanding of how the interactions between the components change, producing structural changes in the dynamics of the system.

The first step is to define the boundaries of the system that is to be the subject of the study; and to determine its components, how these components interact with each other and with variables outside the boundaries that have been defined. As Garcia (1981, 1986) points out, the process of system definition continues throughout the life of a project, as new processes are found to have an important role in the dynamics of the system.

In order to understand the spatial structure and dynamics of the effects of human activity, a scheme for describing a region in terms of landscapes is proposed. It is based on the concept of the *landscape response units* (LRU) (Ramia, 1980; Perez-Trejo, 1988). Figure 7.1 illustrates how the LRU can be conceptualized as the intersection of three process oriented components: (1) the physical processes, including climate; (2) the ecotrophic module (ETM) (Cousins, 1990) described in Chapter 4; and (3) the human-economic-generated processes.

The LRU framework in Figure 7.1 shows how spatial patterns in the landscape affect and are affected by not only abiotic processes such as water flows, floods, nutrient flows, and erosion (Lopes *et al.*, 1977; Ramia, 1980; Imeson, 1987; Gerits *et al.*, 1987), but also biotic processes that make up the ETM, such as movement of individuals in populations, migration and recruitment rates (Ambuel and Temple,

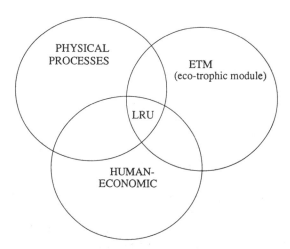

Figure 7.1. Process-oriented representation of the components of LRUs.

1983; Schoener and Schoener, 1983; Urban and Shugart, 1985), and by human processes of cultivation, extraction, deforestation, grazing, and water management. The
LRU is defined in terms of characteristics of the spatial ecology of the system each
landscape described in terms of a geographic location covering several square kilometres, geomorphology, soils, vegetation, and response to intervention (such as
burning, flooding or grazing).

The perception of the evolution of a landscape

To add to the difficulty of understanding the complex dynamics of landscapes, landscape ecologists must contend with the complication of the immense spatial and
temporal scales of natural systems. In trying to study and manage this overwhelm-
ing complexity of the spatial and temporal dynamics, it is both natural and necessary to be led to organize, categorize and aggregate into patterns, communities and
populations which are then characterized in terms of some average behaviour. The
first task of mapping these features from satellite images into thematic maps is the
subject of Chapter 11 in this volume. The difficulty stems from the large amount of
variability and noise that each pixel contains. Any feature of the landscape, such as
vegetation cover, soil type or hydrological status produces spectral properties which
make it difficult to classify with certainty into any one class. This is particularly
problematic in transition zones or ecotonal areas. Hole and Campbell (1985) point
out the problems of generalization errors that can occur in the boundary region of
soil patterns and can generate serious classification problems.

Advanced techniques such as those described in Chapter 11 can help to elimi-
nate the errors in classification in this first phase of producing thematic maps. The
next phase requires understanding of how the different classes of land types in the
thematic map relate to each other spatially and temporally. Each boundary in the
thematic map represents areas of flows of materials, water, nutrients and information between the landscape elements that have been characterized. It is the flows
that maintain the boundary where it is, and it is these flows that determine where
the boundary might be shifting to over time. Each landscape unit is characterized in
terms of variables of interest, such as: cover, productivity, water-holding capacity,
concentration of nutrients, soil characteristics, and wildlife sightings. All the pixels
representing a land class are assigned values for the characteristics that are of interest, as a means of determining the behaviour of the different land units in the landscape. This characterization of the landscape ecosystems is represented in terms of
average values obtained from field surveys and detailed site studies. Even though it
is a necessary simplification, it leads to a characterization of the system in what
Allen and McGlade (1986) termed as 'probabilistic in character', reducing the richness of the system's diversity to the most likely, average behaviour, which produces
'machine-like' dynamics into which change can only be introduced as an exogenous
driving force.

GIS techniques can be very useful in exploring the spatial distribution of the
variables that might play a role in understanding ecosystem dynamics. For example,
Agee *et al.* (1989) showed how whitebark pine, subalpine larch and subalpine herb
cover was 85% accurate in associating 91 recorded grizzly bear sightings in the
Northern Cascades of Washington with these cover types. These results indicate
that community characteristics can be used to predict where grizzly bears might be
sighted. They provide not only a useful management tool for identifying the protec-

tion of potential grizzly bear habitat using GIS, but also provide clues about habitat characteristics required by the bears. What is more difficult is understanding how these observed preferences might change—How much does bear activity actually affect these changes, or how different would the results be if the same study were carried out in another region, such as Yellowstone Park? These are questions that need to be addressed in a landscape ecology context in order to understand the linkages between components of the ecosystem that could generate more general ideas about the way that bears interact with their environment.

The processes that shape the landscape, such as Schumm's (1977) non-linear dynamics of sedimentation and erosion in riverbeds, occur as a result of biological, physical and climatic interactions generating dynamics where communities become established in a landscape. They tend to have an effect on the concentration and reorganization of the physical weathering processes, and to change the physical and chemical nature of soils, increasing infiltration rates and accumulating organic matter. The changes in these processes eventually reach thresholds where the local dynamics generate global effects and bring about a restructuring of the communities, thus creating new landscape-level rates of erosion and weathering.

Schumm (1977) illustrates these non-linear processes when he describes the complex evolution of drainage basins (Figure 7.2). When the alluvial depositions of basins are examined, the non-uniformity and variability of the depositions within and across basins makes it impossible to explain errosion and deposition events based on exogenous events such as climate change in the last 10 000 years. An initial incision in a channel may be due to biological or human activity, such as climate change or extensive deforestation, but then the enormous sediment loads that are produced cause rapid deposition and an increase in gradient upstream. This deposition forms an alluvial plain creating positive feedback for more sediments to be deposited, until the gradient reaches a threshold level and the alluvial deposition erodes rapidly as the sediment load decreases.

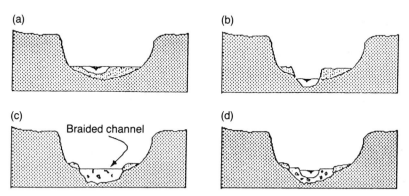

Figure 7.2. An illustration of Shumm's (1977) non-linear dynamics of sedimentation and erosion in riverbeds shows an event (climate change or land-use) causing an erosional response that occurs initially at the mouth of the drainage basis. The change in gradient upstream moves the erosion progressively upstream, incising the main channel to form a terrace (a, b). As erosion progresses upstream, sediments increase in the main channel, resulting in depositions and formation of a braided stream (c). As sediment yields decrease, there is a renewed incision of the alluvial deposits to form low terraces (d). Because the events that trigger these complex spatial patterns of erosion and sedimentation can be distributed over vast areas of a drainage basin, there can be many events happening at different times, preventing the fluvial system from reaching equilibrium.

Table 7.1. *Landscapes of the Alto Apure savannas.*

Landscape	Landscape response unit	%	Soil
Ancient savanna (Pleistocene)	Medano	9.6	Sandy
	Silty-bajio	10.6	Silty
	Sandy-bajio	49.7	Sandy–clay
	Banco	10.0	Sandy–loam
	Loam-bajio	20.1	
Recent sedimentary savanna (Holocene)	Banco	14	Sandy–loam
	Bajio	65	Clay–loam
	Estero	20	Clay
	Medano	1	Sandy

As the physical forces of erosion tend to shape a landscape, vegetation and animal communities explore and colonize those sites that have sufficient moisture, or nutrients for them to become established. These communities begin to change the site's characteristics and generate a pattern of LRUs that interact to determine the global rates of erosion, run-off, and sedimentation patterns. For example, in the flooded Llanos of Venezuela the differential deposition of sediments initiates the formation of the sandy ridges (Bajios). As sands accumulate, grasses become established and accelerate the deposition of sediments, creating a natural dam that causes floods for the most part of the year (the Estero). This flooded swale accumulates the finer clay sediments, increasing its water-retaining capabilities, and generating a water-loving community of plants and animals (Ramia, 1980; Perez-Trejo, 1988).

The extent and duration of periodic flooding in these savannas is mainly determined by the spatial patterns of the different eco-physiographic units in the region (Table 7.1). This is a vivid example of the nature of complex evolving systems: the

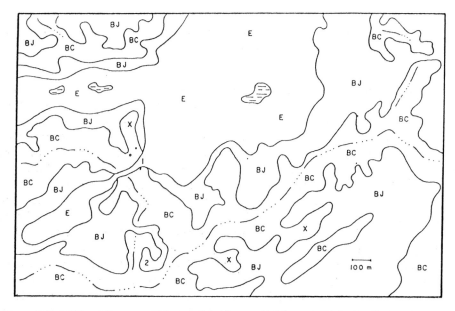

Figure 7.3. *LRUs of the recent (Mantecal) landscape. BC, banco; BJ, bajio; E, estero.* (Adapted from Ramia (1980).)

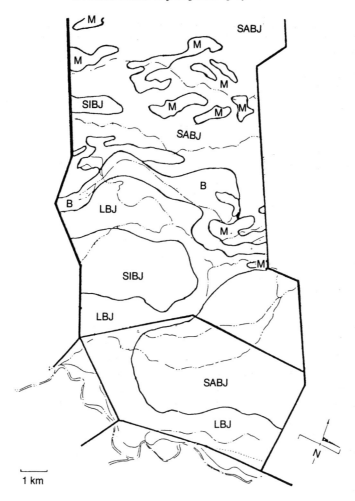

Figure 7.4. LRUs of the ancient (Chichiteras) landscape. LBJ, loam–bajio; SIBJ, silty–bajio; SABJ, sandy baijo; M, medano; B, banco.

extent and duration of flooding is an emergent property of the system, within which the dynamics of each local community takes place. But it is the spatial distribution of those communities which generates the global flooding patterns of the system.

By mapping the spatial pattern over the landscape it was possible to understand how the extent and distribution of LRUs could account for the range and duration of flooding at the landscape level. The spatial patterns shown in Figure 7.3 indicate how the LRUs in the recent savanna are distributed in a mosaic of small (mostly less that 1 km²) patches. In contrast, the spatial patterns of the LRUs of the ancient savanna landscape (Figure 7.4) are much more homogeneously distributed, and cover areas of several square kilometres. Figure 7.5 shows the location of these landscapes in Venezuela. These differences in the spatial pattern of LRUs produce quite different patterns of flooding in the two landscapes. In the recent savannas flooding depths are much more variable over the landscape, varying from 0 cm to above 100 cm over distances of a few hundred metres. In the ancient savanna, flooding levels are much less variable over the landscape, covering about 50% of the ground with a sheet of water of 50–100 cm for 3–4 months of the year.

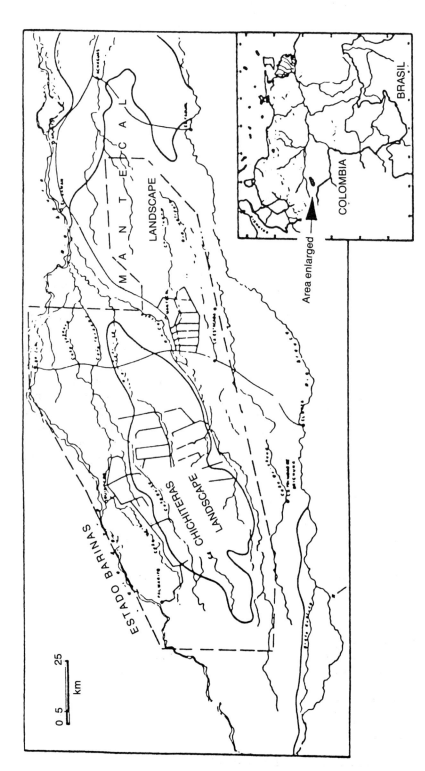

Figure 7.5. The location of the Mantecal and Chichiteras landscapes in the western savannas of Venezuela. (Adapted from Ramia (1980).)

Table 7.2. *Effect of regulated flooding on the Bajio of the recent savanna.*

Natural conditions	Diked flooding
Soil	
Clay–loam, saturated	Clay–loam, flooded
Vegetation (No. of species)	
37	12
Dominant species	
Panicum laxum	*Hymenachne amplex.*
Leersia hexandra	*Leersia hexandra*
Paspalum chaff.	
Axonopus comp.	
Above-ground primary production (t ha^{-1})	
7 ± 1	10

Two LRUs were chosen to exemplify the differences observed in response to flooding. Tables 7.2 and 7.3 the different variables that were used in comparing the response to regulated flooding by the dikes of the 'Bajio' unit in the recent savanna, versus the 'silty Bajio' in the ancient savanna. In both cases there was a significant decrease in the diversity of the plant species that invade under flooded conditions, both in terms of the total number of species and in terms of the species that account for most of the above-ground biomass. The response in total above-ground biomass due to regulated flooding was marked by a significant increase in the Bajio of the recent savanna, in contrast to the very slight increase in the above-ground biomass of the silty Bajio of the ancient savanna. The differences in the productivity of the two landscapes can be explained in terms of the relatively lower levels of soil nutrients found in the ancient savanna (Ramia, 1980), and probably in terms of the differences between the two landscapes in the effect that spatial patterns have on the dynamics of nutrient flows.

Table 7.3. *Effect of regulated flooding on the Silty-Bajio of the ancient savanna.*

Natural conditions	Diked flooding
Soils	
Silty, saturated	Silty, flooded
Vegetation (No. of species)	
20	10
Dominant species	
Mesosetium chasae	*Paratheria prostata*
Axonopus anceps	*Leersia haexandra*
Andropogon brevifolius	
Axonopus purpusii	
Sorghastrum parvifl.	
Leersia hexandra	
Bare soil 35%	
Above-ground primary production (t ha^{-1})	
2.5	3.5

Discussion

The central idea of this chapter is that landscapes are part of complex natural systems that can be characterized by three major features which determine the context in which a methodological framework can be developed. Such systems are the product of an evolutionary process, they display spatial structure that is self-generated (at least in part), and they possess non-linear dynamics that explain the unpredictable nature of the response to man's accumulated impacts. Because these systems are the product of an on-going evolutionary process, their dynamics follow trajectories that are far from equilibrium. The system components are therefore the product of a long process of mutual adjustment and coevolution, and they are characterized by non-linear responses to changes in the environment, or as a result of human intervention.

The interaction between the biotic, the abiotic, and the human processes is what creates and maintains the structure and flows in the landscape. Plant and animal populations are dealing simultaneously with problems of obtaining their requirements for growth and reproduction (sunlight, nutrients and water) whilst being embedded within an interconnected network of interacting populations. At a local level, physical and mechanical forces tend to shape certain types of soil and sediment into units with specific characteristics (soil texture, nutrient contents and water retention). As plants and animals colonize each of these units they tend to change some of these characteristics. As organic matter accumulates, the water-holding capacity may increase. The initial conditions of the site and the history of the community will determine the direction and rate of change of the characteristics in each unit. The members of the present community of a site may change these characteristics just enough to make the site better suited for other populations to become established. These interactions go on within and between populations, generating some spatial pattern that might be recognized and used to characterize the landscape in terms of LRUs.

It is evident that social and cultural factors as well as government policies and subsidies influence land-use practices. What are not well understood are the mechanisms and causal links that govern the interactions among these physical, climatic, biological and human factors that might lead to changes in the patterns of the landscape and even land degradation. For example, the land degradation occurring in many regions can be attributed to changes in agricultural practices. The impact of these practices has been further aggravated by government subsidies that provide an economic incentive which makes these new land-use practices much more attractive. The ecological and environmental responses that are being observed result from the complex interactions between changing value systems and economic incentives on the part of farmers and national or international policies.

It is the dialogue between local dynamics and global patterns that produce the seemingly stable spatial patterns of the landscape. The spatial patterns that are observed in a system cannot be explained in terms of exogeneous changes, in fact they are a natural consequence of the biotic–abiotic interactions of the components in the system. For example, fire is a part of the ecology of many landscapes. If it was conceived as a 'disturbance' to some current equilibrium, it would imply that the equilibrium communities in the landscape would resist its impact. But, inevitably, the impact of fire would lead to a complete denudation of the landscape, reduction of diversity, and eventual degradation. In studying landscapes that have been subjected to fire in their evolution, what is observed is the colonization and gradual adaptation of soil organisms and plant communities that 'learn' to coexist with periodic fire.

The LRU concept can be a useful way of gaining further understanding of the complex linkages between the physical, ecological, climatic and human components of the landscape ecosystem. Overlaying techniques are a powerful tool for visualizing these interactions (Burrough, 1986; Goulter and Forrest, 1987; Bailey, 1988; Agee *et al.*, 1989). However, these techniques could benefit from considering landscape flows and structures that make up the LRU. In this way the difficult task of combining different types of spatial information for gaining further understanding of the complex dynamics of the landscape may be achieved.

References

Agee, J. K., Stitt, S. C., Nyquist, M. and Root, R., 1989, A geographic analysis of historical grizzly bear sightings in the North Cascades, *Photogrammetric Engineering and Remote Sensing*, **55**(11), 1637–1642.

Allen, P. M., 1988, Evolution: Why the whole is greater than the sum of the parts, *Ecodynamics, Contributions to Theoretical Ecology*, 2–30.

Allen, P. M. and McGlade, J. M., 1986, Optimality, adequacy and the evolution of complexity, *Proc. MIDIT Conference*, Copenhagen, August 1986.

Allen, P. M., Sanglier, M., Engelen, G. and Boon, F., 1985, Towards a new synthesis in the modelling of evolving, complex systems, *Environment and Planning B: Planning and Design*, **12**, 65–84.

Ambuel, B. and Temple, S. A., 1983, Area-dependent changes in the bird communities and vegetation of southern Wisconsin forests, *Ecology*, **64**, 1057–1068.

Bailey, R. G., 1988, Problems with using overlay mapping for planning and their implications for geographic information systems, *Environmental Management*, **12**(1), 11–17.

Burrough, P. A., 1986, *Principles of Geographic Information Systems for Land Resources Assessment*, Oxford: Clarendon Press.

Cousins, S. H., 1990, Countable ecosystems deriving from a new food web entity, *Oikos*, **57**(2), 270–275.

Forman, R. T. T. and Godron, M., 1986, *Landscape Ecology*, New York: Wiley.

Garcia, R. V., 1981, *Drought amd Man, Vol. 1: Nature Pleads Not Guilty*, Oxford: Pergamon Press.

Garcia, R. V., 1986, Conceptos basicos para el estudio de sistemas complejos, in Leff, E. (Ed.), *Los Problemas del Conocimiento y la Perspectiva Ambiental del Desarrollo*, Mexico: Siglo 21.

Gerits, J., Imeson, A. C., Verstraten, J. M. and Bryan, R. B., 1987, Rill development and badland regolith properties, in Bryan, R. B. (Ed.), *Rill Erosion*, West Germany: CATENA-Verlag.

Goulter, I. C. and Forrest, D., 1987, Use of geographic information systems (GIS) in river basin management, *Water Science Technology*, **19**(9), 81–86.

Hole, F. D. and Campbell, J. B., 1985, *Soil Landscape Analysis*, Totowa, NJ: Rowman and Allenheld.

Imeson, A. C., 1987, Soil erosion and conservation, in Gregory, K. J. and Walling, D. E. (Eds), *Human Activity and Environmental Processes*, New York: Wiley.

Lopes, D., Tugues, J. L., Bulla, L. and Briceño, M., 1977, Balance nutricional en un ecosistema de sabana inundable, *IV Simposium Internacional de Ecologia Tropical 2*, Panama, pp. 645–660.

Naveh, Z. and Lieberman, A. S. (Eds), 1984, *Landscape Ecology: Theory and Application*, New York: Springer-Verlag.

Nicolis, G. and Prigogine, I., 1977, *Self-Organization in Non-Equilibrium Systems*, New York: Wiley Interscience.

Perez-Trejo, F., 1988, Impact assessment methodologies for complex natural systems, *Proc. International Association for Impact Assessment*, The Netherlands, June.

Prigogine, I. and Stengers, I., 1987, *Order Out of Chaos*, New York: Bantam Books.

Prigogine, I., Allen, P. M. and Herman, R., 1977, The evolution of complexity and the laws of nature, in Lazlo, E. and Bierman, (Eds), *Goals in a Global Community*, New York: Pergamon Press.

Ramia, M., 1980, Relaciones geomorfológicas-suelo-vegetación en el Alto Apure, *Trabajo de Ascenso*, University of Central Venezuela.

Schoener, T. M. and Schoener, A., 1983, Distribution of vertebrates on some very small islands and patterns in species numbers, *Journal of Animal Ecology*, **52**, 237–262.

Schumm, S. A., 1977, *The Fluvial System*, New York: Wiley.

Urban, D. L. and Shugart, H. H., 1985, Avian demography in mosaic landscapes, in *Wildlife 2000: Modelling Habitat Relationships for Terrestrial Vertebrates*, Madison, WI: University of Wisconsin Press.

Urban, D. L., O'Neill, R. V. and Shugart Jr, H. H., 1987, Landscape ecology, *Bioscience*, **37**, 119–127.

PART IV

Techniques and technical issues

8

Problems of sampling the landscape

A. R. Harrison and R. Dunn

Introduction

The complex spatial nature of the landscape presents a number of problems when designing optimal sampling schemes for surveys of land use. First, there is a lack of information on the spatial characteristics of the landscape. Second, little attention has been given to extensions of traditional sampling theory for two-dimensional sampling of land use. Here we present a methodology which uses a digital represen-tation of the landscape derived from classified Landsat thematic mapper (TM) imagery to provide spatial data on land-use characteristics, which in turn provides the basis for a series of simulation experiments which tested the efficiency of differ-ent sampling approaches.

The work was carried out under a contract placed by the Department of the Environment to assist in the planning of a repeat national survey of land use. In March 1984, the Department of the Environment and the Countryside Commission jointly commissioned a 2-year project to monitor changes in the landscape of England and Wales. The principal objective of this monitoring landscape change (MLC) project was to provide quantitative data on changes in the distribution and extent of major landscape features beginning with the late 1940s to the present (Hunting Technical Services Ltd, 1986). The key components of the MLC survey methodology were area sampling and airphoto interpretation (API). We have analysed the sampling procedures employed in the MLC project and have demon-strated two related statistical drawbacks: too few observations and a very variable intensity of sampling (Harrison et al., 1989). Our recommendation that a repeat survey adopt a new sampling approach is partly based on the simulation work presented here.

It is assumed throughout that area sampling is used to measure percentage cover of land use, rather than point or line (transect) sampling. This is in accord with most recent practice in recent sample-based land-use surveys in the UK (Hunting Technical Services Ltd, 1986; Bunce et al., 1981; Nature Conservancy Council, 1987).

Problems of sampling the landscape

Our analysis of the MLC data, and wider considerations, suggest a number of prob-lems relating to the use of traditional sampling theory when applied to a two-dimensional natural population such as the landscape. First, most sampling theory assumes either a population of discrete elements, from which a sample is taken, or a continuously varying function, from which measurements are taken. However, for studies of land use the population is the landscape which has a much more complex nature. Second, when sampling the landscape a number of attributes (i.e. the different land-use categories) are usually measured within the sample site. This multivariate nature of the data makes defining an optimal sampling scheme difficult, since what may be appropriate for one category may be less optimal for others.

A related problem is the extent to which an appeal to traditional sampling theory can determine optimal sample site size. This is a central component of sampling schemes for the measurement of land use, but traditional statistics and work on spatial sampling has little to say on the question. Our analysis of the MLC data suggests that the influence of sample site size on estimates of land use can be considerable and may constitute the largest measurement error in the data set (Harrison *et al.*, 1989). Here the complex nature of the landscape makes deciding upon an optimum size of sample site difficult as it will depend upon a knowledge of the spatial configuration of land use at each location.

Taken together, these problems question the way traditional sampling theory should be used when designing a sampling scheme to measure land use. This is a neglected area of research and the statistics and spatial sampling literature have little to offer in the way of guidance. As a result there is a lack of accumulated knowledge of the spatial characteristics of the landscape, and insufficient consideration has been given to the necessary extensions of two-dimensional sampling theory for land-use studies.

One factor explaining this relative lack of attention has been the difficulty of obtaining complete enumerations of spatial populations. For land-use studies the availability now of fine spatial resolution satellite imagery provides a potentially rich data source for analysis of the spatial properties of the landscape. Here we present a methodology which uses a digital representation of the landscape derived from Landsat TM satellite imagery to provide population data on land-use characteristics, which in turn provide the basis for a series of simulation experiments into aspects of sampling land use.

Data and method

A subscene corresponding to the County of Avon, UK, was extracted from a Landsat TM scene (path/row, 203/024; date, 24 April 1984). The subscene was registered to the UK National Grid using seven control points from Ordnance Survey 1 : 50 000 maps and resampled using nearest-neighbour interpolation.

A supervised maximum likelihood classification using Landsat TM spectral bands 3, 4, 5 and 7 was carried out using training data for four broad land-use categories: arable, grassland, urban and woodland. Training areas (typically 4–6) were selected for each land-use category based on information derived from 1 : 50 000 maps, API, and image tone, context and pattern. The procedure resulted in a classified image in which each pixel was allocated to one of the four land-cover types or to an unclassified category. A number of comparisons between the Landsat TM classifications and land-use statistics for Avon indicated a satisfactory level of classification accuracy.

Treating the classified Landsat TM image as a realistic representation of landscape pattern we then investigated a number of sampling strategies. For areas of 16,

8, 4, 2, 1 and 0.5 km² we defined a square grid on the image. This grid thus divided the image completely into non-overlapping squares of the appropriate size. Within each square we then calculated and recorded the percentage cover of land use for each of the four land-use categories in the classified image. Only those squares which lie completely in the County of Avon were included in the analysis.

Statistics of simulation: mean cover and variability

Table 8.1 summarizes some features of this sampling process. In particular, the final column of the table shows some important edge effects. These edge effects may introduce bias where land use is more or less concentrated near the edge of Avon. It is worth noting that this approach is only one of many possible ways of superimposing a grid on the 'landscape', i.e. the image. We could have chosen a set of different starting positions for the grid which would have generated considerably more data. In addition, we might have restricted our analysis to that area covered by th 16 km² sites, but that would have reduced the sample size for smaller sites.

Table 8.2 presents the estimate of the mean percentage cover of the four land-use categories for the different sizes of sample site. The estimates for woodland and urban show some trends: for woodland the estimate increases as site size decreases, whereas the pattern for urban areas is the reverse. These trends appear to be due to the spatial pattern of these land uses in Avon. Much of the woodland is near the periphery of the county, so that it stands a higher chance of being omitted when the size of sample site is large. In contrast, urban land use is largely concentrated in the centre of the county so that when the edges are 'lost' the remaining land has a higher proportion of urban cover. The estimates for grassland and arable land show no such trend, reflecting the more homogeneous spread of these categories throughout Avon.

Table 8.1. Summary statistics of simulation.

Site size (km²)	No. of complete sites	Total area (km²)	Area as % of Avon
16	59	944	70
8	133	1064	79
4	282	1136	84
2	605	1210	90
1	1237	1237	92
0.5	2497	1249	93

Table 8.2. Means.

Site size (km²)	Woodland	Grassland	Arable	Urban
16	2.98	29.69	27.83	14.98
8	3.07	29.22	28.42	14.23
4	3.01	29.03	28.81	14.06
2	3.15	29.11	28.20	13.91
1	3.13	28.98	28.50	13.77
0.5	3.18	28.93	28.17	13.79

Table 8.3. Standard deviations.

Site size (km²)	Woodland	Grassland	Arable	Urban
16	4.16	9.85	13.97	16.75
8	5.08	10.95	15.71	17.58
4	5.19	12.36	17.13	18.37
2	6.35	14.13	18.43	18.94
1	6.97	16.02	20.27	20.15
0.5	7.98	18.10	22.08	21.18

Table 8.3 presents the standard deviation as an index of variability for each of the four land-use categories at different site sizes. Two points emerge from this table. First, the standard deviation increases as the site size falls. This is not unexpected as smaller sites are likely to give more variable results as more extreme (larger or smaller) percentages of use in one particular category become more likely. Second, the values of the standard deviations appear to reflect the way these land classes cluster on the ground. Thus urban and arable land cluster more strongly compared with the other two categories. The spatial configuration of the land-use categories thus emerges in this variability (for details see Harrison *et al.*, 1989).

Simulating sampling schemes: size of site

By treating the characteristics of the simulated sampling process (e.g. number of observations (n) and variability (σ)) as 'population parameters' it was possible to derive information on estimated sampling distributions for different random sampling schemes. Assuming a random sampling scheme, for a given sample size and for a particular sample site size, we can calculate the expected standard deviation of the sample mean ($\tilde{\sigma}$) using the formula: $\tilde{\sigma} = \sigma/\sqrt{n}$.

There are clearly numerous scenarios which we could investigate with this information. Here we focus on one scenario. We took samples which covered 64 km² of the land surface of Avon in total, about 5% of the land surface (much in line with the MLC sampling scheme used in Avon). Given that the size (area extent) of the sample sites used may vary, this can be done in a number of ways: four 16-km² sites, eight 8-km² sites, and so on, down to 128 0.5-km² sites.

Table 8.4. Random sampling with 64 km² of cover, standard error of the sample means.

Sample					
No.	km²	Woodland	Grassland	Arable	Urban
4	16	2.08	4.93	6.99	8.38
8	8	1.80	3.87	5.55	6.22
16	4	1.30	3.09	4.28	4.59
32	2	1.12	2.50	3.26	3.35
64	1	0.87	2.00	2.53	2.52
128	0.5	0.71	1.60	1.95	1.87

Table 8.4 reports the expected standard errors of the sample means under this experiment for the four land-use categories, based on the standard formula given above. These results show that, for a given total area of land use, the most efficient sampling approach is to take many small sites. Although this result is perhaps not unexpected, these results do allow us to quantify the expected differences in efficiency of different sampling strategies. Thus using the results in Table 8.4, if we moved from a strategy of 16 4-km^2 sites to 64 1-km^2 sites we would expect to reduce the standard error of the means by 33% for woodland, 35% for grassland, 41% for arable land, and 45% for urban land. The consistency of the results from this experiment suggests that there are very substantial gains to be had from moving to the use of more smaller sites for sampling land-use cover.

Simulating sampling schemes: random versus systematic

A further important question we can address with these data is the relative efficiency of different sampling strategies. Here we compare random, stratified random (with one observation per stratum) and systematic strategies.

To evaluate systematic sampling, a square grid was superimposed over the data such that the spacing of the grid was an integer multiple r of the side length of the sample sites. Odd values of r were used in the experiments since the central site within each grid square is sampled. For any value of r the origin of the grid may be in one of r^2 positions so that an empirical sampling distribution for a systematic scheme may be obtained. The variance of the sample means derived in this way are written V_{sys}. Here values of $r = 5, 7, 9$ and 11 were used, yielding median sample sizes of 100, 51, 31 and 21, respectively. Due to edge effects, the exact size of the sample varied between the r^2 replications.

For stratified random sampling a grid of the same size as for systematic sampling was used, but in this case a site was chosen at random within each grid square. For comparability with systematic sampling the distribution of the sample mean was determined from r^2 replications; the variance of this distribution is written as V_{st1}. The sampling distribution for random sampling was also determined empirically; its variance is denoted by V_{ran}.

The values V_{sys}, V_{st1} and V_{ran} were calculated for the four land-use categories for grid sizes $r = 5, 7, 9$ and 11. Table 8.5 shows the gain in efficiency of moving from random sampling to systematic sampling by recording the ratio V_{ran}/V_{sys} for each

Table 8.5. Comparison (V_{ran}/V_{sys}) of the variances of the sample mean from systematic sampling (V_{sys}) and random sampling (V_{ran}) for different sampling intensities.

Land cover	r			
	5	7	9	11
Grassland	1.82	1.93	0.98	1.61
Arable	1.39	1.67	2.00	1.87
Urban	3.85	4.21	1.51	1.52
Woodland	2.24	1.11	1.47	0.98

value of r. These experimental results show that, on average, very large gains in
efficiency result from moving to systematic sampling from random sampling, with
typical values between 67% and 37%. These values are similar to those found in
previous studies, which report comparable statistics (Matern, 1960; Payandeh,
1970). The pattern of average values also suggests that larger gains in efficiency tend
to occur at greater sampling intensities (smaller values of r).

   The individual entries in Table 8.5 show great variability both between land
cover types and for the same cover type at different sampling intensities. For
example, urban land has the largest gains in efficiency at $r = 5$ and 7, but only just
above average gains at $r = 9$ and 11. Two possible explanations for these highly
variable results may be forwarded. One is that they are due to small sample effects
within the simulation exercises. The second is that they reflect actual differences in
the spatial configuration of the land-use types.

   To investigate the latter hypothesis, spatial autocorrelation functions were calculated for the four land-use types to see if this additional information assisted in
the interpretation of Table 8.5. The lag interval used is equivalent to the length of
the sample sites (0.71 km), and in each case the autocorrelation function was calculated for the east–west and north–south directions, to a maximum of 15 lags. The
results are shown in Figure 8.1.

   For improved grassland and arable land the autocorrelation functions are close
to a geometric decline, the model assumed in much theoretical work. For urban

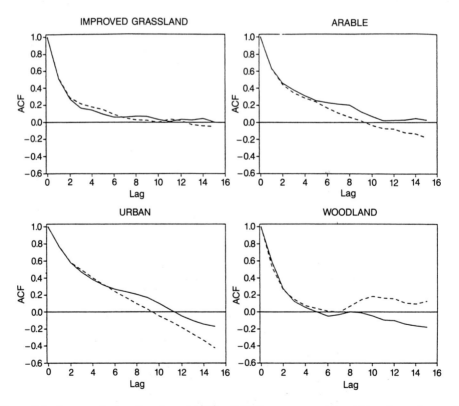

*Figure 8.1. Autocorrelation functions (ACF) of land-use categories within Avon. (– – – –)
North–south direction; (———) east–west direction. Lag interval 0.71 km.*

land the negative correlations at longer lags suggest a very strong degree of association across the study area. Reference to the original data shows this to be the case, since most urban land is concentrated in the large city areas of Bristol and Bath. Finally, the pattern for woodland suggests the possibility of some periodic variation at around lag 10 in the north–south direction.

In some cases these results assist in the examination of Table 8.5: the high degree of correlation for urban land is matched by very large gains in efficiency for that category; the poor performance of systematic sampling of woodland at $r = 11$ may be due to periodic variation. In other cases the correspondence is less helpful. For example, the poor performance of systematic sampling at $r = 9$ for improved grassland is unexplained. In summary, the autocorrelation functions do assist a fuller interpretation of the results of Table 8.5, but certain aspects remain unclear. There is some suggestion that unusual autocorrelation functions may result from particular spatial configurations, which in turn affect the efficiency of systematic sampling. On a wider note, these empirical results indicate that autocorrelation functions of natural populations may take quite complex forms. To assume a geometric form of decline may be appropriate in some cases, but not all.

To complete the analysis of alternate sampling schemes Tables 8.6 and 8.7 present the ratios V_{ran}/V_{st1} and V_{st1}/V_{sys}, to assess the relative performance of stratified random sampling compared to random and systematic sampling. The gain in efficiency of moving from random sampling to stratified random sampling (V_{ran}/V_{st1}) shows a similar pattern to the data in Table 8.5: there are, on average, considerable gains in efficiency, although these are less than for systematic sampling. The gain in

Table 8.6. Comparison (V_{ran}/V_{st1}) of the variances of the sample mean from random sampling (V_{ran}) and stratified sampling (V_{st1}) for different sampling intensities.

Land cover	r			
	5	7	9	11
Grassland	1.31	1.37	0.92	1.03
Arable	2.14	1.41	1.34	1.26
Urban	2.91	2.48	1.68	1.31
Woodland	1.16	1.53	1.67	1.23

Table 8.7. Comparison (V_{st1}/V_{sys}) of the variances of the sample mean from stratified sampling (V_{st1}) and systematic sampling (V_{sys}) for different sampling intensities.

Land cover	r			
	5	7	9	11
Grassland	1.38	1.42	1.06	1.56
Arable	0.64	1.18	1.49	1.49
Urban	1.32	1.70	0.90	1.16
Woodland	1.94	0.73	0.88	0.98

efficiency of moving from stratified random sampling to systematic (V_{st1}/V_{sys}) is generally small, but the average ratio of between 17% and 3% suggests the gain is worth having.

The results reported here illustrate both the advantages of a systematic approach and that certain important unresolved issues exist, many of which relate to the complexity of the two-dimensional case. In accord with previous theoretical and empirical studies (Cochran, 1946; Finney, 1948; Osbourne, 1942; Williams, 1956), systematic sampling consistently outperforms alternative strategies for the natural population studied here. But the gain in efficiency is highly variable, and appears to reflect the complex and varied autocorrelation functions of the data.

Conclusions

Four broad conclusions arise from these simulation experiments. First, satellite imagery represents a valuable resource for experimental approaches to landscape simulation and for testing the efficiency of different sampling approaches.

Second, moving to using more smaller sites to measure land use should increase the accuracy of measurement compared to an equivalent area of land in larger sites.

Third, in general, systematic sampling provides the most efficient sampling approach. Given the operational advantage of the design, its use in sample-based land-use surveys has much to recommend it. (However, there may be problems with this approach if periodic variation is present in the landscape in which case a stratified random approach is preferable.)

Fourth, the gains in efficiency are highly variable, being a function of both land-use type and sampling intensity. In part this appears to be due to the complex nature of spatial dependence in the natural populations studied here. Whether such patterns are widespread can only be ascertained by further empirical studies.

Acknowledgement

Part of this work was carried out under contract PECD 7/2/47 placed by the Department of the Environment. The views expressed here are not necessarily those of the Department of the Environment or any other Government department.

References

Bunce, R. G. H., Barr, C. J. and Whittaker, H. A., 1981, Land classes in Great Britain: preliminary descriptions for users of the Merlewood method of land classification, *Institute of Terrestrial Ecology, Merlewood Research and Development Paper No. 86.*

Cochran, W. G., 1946, Relative accuracy of systematic and stratified random samples for a certain class of population, *Annals of Mathematical Statistics*, **17**, 161–177.

Finney, D. J., 1948, Random and systematic sampling in timber surveys, *Forestry*, **22**, 64–99.

Harrison, A. R., Dunn, R. and White, J. C., 1989, A statistical and graphical examination of monitoring landscape change data, *Final Report to the Department of the Environment, Research Contract PECD 7/2/47*, 3 Vols, London: DoE.

Hunting Technical Services Ltd, 1986, Monitoring landscape change, *Final Report to the Department of the Environment*, 10 Vols, London: DoE.

Matern, B., 1960, Spatial variation, *Meddelanden Fran Statens Skogsforsknings Institut*, **49**(5), 1–144.

Nature Conservancy Council, 1987, Research and survey in nature conservation, No. 6, Changes in the Cumbrian countryside, *First Report of the National Countryside Monitoring Scheme*, Peterborough: NCC.

Osborne, J. G., 1942, Sampling errors of systematic and random surveys of cover-type areas, *Journal of the American Statistical Association*, **37**, 256–264.

Payandeh, B., 1970, Relative efficiency of two-dimensional systematic sampling, *Forestry Science*, **16**, 271–276.

Williams, R. M., 1956, The variance of the mean of systematic samples, *Biometrika*, **43**, 137–148.

9

A methodology for acquiring information on vegetation succession from remotely sensed imagery

D. R. Green, R. Cummins, R. Wright and J. Miles

Introduction

Traditional approaches used by ecologists to examine, understand and map patterns of natural or semi-natural vegetation have frequently been constrained by spatial, temporal and economic problems. These problems have been particularly serious in studies of successional changes where long time-scales may be involved. Some changes in upland vegetation, for example, may take decades before they are detectable (Ball *et al.*, 1982). However, a variety of remote-sensing techniques offer the ecologist a potential means to overcome such problems. In particular, extensive and frequent use has been made of multi-temporal panchromatic aerial photographs, at a variety of scales, to classify and map vegetation (Fuller, 1981). More recently, satellite data and imagery have been used to identify upland vegetation types (Hume *et al.*, 1986; McMorrow and Hume 1986; Morton, 1986; Jewell and Brown, 1987; Weaver, 1987).

However, most studies have been concerned with survey-type work of large-scale changes in land use such as agriculture and forestry. In contrast, relatively little attention has been paid to natural or semi-natural habitats (Wyatt, 1984) and, despite the potential of remote sensing for collecting information on vegetation successions, until recently there have been few methodologies which permit the statistical analysis of such information. It is only the recent advances in computing power that have enabled the manipulation and analysis of the large data sets arising from these (necessarily) high-definition studies.

Our knowledge of the direction, rates and frequencies of different successional pathways of vegetation change in Britain has advanced only slowly and the shortage of relevant data severely limits the construction of even simple mathematical models of successions. Such models are essential if one is to predict with any precision the broad-scale consequences (e.g. at catchment level) of changes in land use and management.

Below, a methodology is outlined which uses remote sensing to provide some data for modelling vegetation successions. It forms a major part of a joint project developed over the past 3 years by the Institute of Terrestrial Ecology (ITE) at Banchory, Grampian Region, and the Department of Geography, Centre for Remote Sensing and Mapping Science, at the University of Aberdeen. The work carried out by ITE is based on 40 0.25 km² samples throughout the north-east of

Scotland, and also examines the changes that occur in soil chemistry arising from changes in vegetation. The University researchers are looking at the vegetation changes in one particular area of major ecological interest in mid-Deeside, Grampian. Data from the latter studies are used to exemplify the sorts of results obtained.

Research objectives

There were three major objectives:

1. to establish a practical methodology for collecting data on vegetation change between 1947 and 1985 in north-east Scotland using high-quality panchromatic aerial photography;
2. to develop a methodology to store the data in a computer-compatible form for ease of access, manipulation, analysis and output, and to establish an ecological database; and
3. to use the data to test and refine existing hypotheses of successional changes in vegetation and to generate new hypotheses where appropriate.

Development of methods and techniques

Figure 9.1 summarizes the overall methodology of the study.

Source of data

Panchromatic aerial photography of high resolution and freedom from cloud cover was selected for 3 years (1947, 1964, and 1985) and obtained from the Scottish Development Department (SDD). Scales of the photography for each date were 1 : 10 000, 1 : 25 000 and 1 : 13 000, respectively.

Study area

The study focused on an area of approximately 20 km × 20 km in the vicinity of the Muir of Dinnet National Nature Reserve in mid-Deeside, Grampian Region, approximately 56 km (35 miles) west of Aberdeen (Figure 9.2). Within this area, eight Ordnance Survey 1 km × 1 km grid squares were identified as 'key' areas most likely to display significant and detectable changes in vegetation succession over time.

The natural/semi-natural vegetation of the area in general is predominantly dry mature heathland. However, birch trees and, to a lesser extent, pine trees are rapidly invading the area. Grazing by wild and domestic animals maintains unimproved grasslands of various sizes and these contain patches of rushes where the drainage is poor. Rush-dominated areas are extensive amongst some of the bogs, and elsewhere form discrete patches in the wetter heaths. Occasional patches of gorse, broom, willow and juniper occur throughout the area. Pockets of bracken are frequent throughout the heather and more extensive areas of bracken often form an understorey in the birch woodland. A full description of the Dinnet Reserve is provided by Marren (1980).

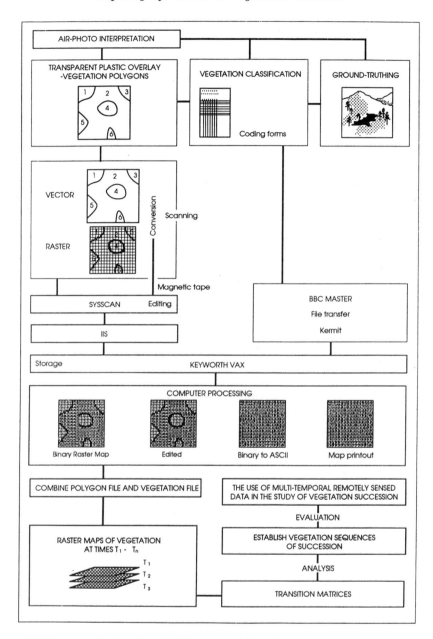

Figure 9.1. Summary of research methodology.

Vegetation classification

The vegetation codes (Table 9.1) were entered either singly or in combination into the appropriate columns on the data sheets (Figure 9.3) to describe the vegetation of each mapped polygon at two levels:

1. as a broad 'overview' in the 'Code' column; and
2. in cover classes of $> 50\%$, 26–50%, 11–25% and $\leq 10\%$.

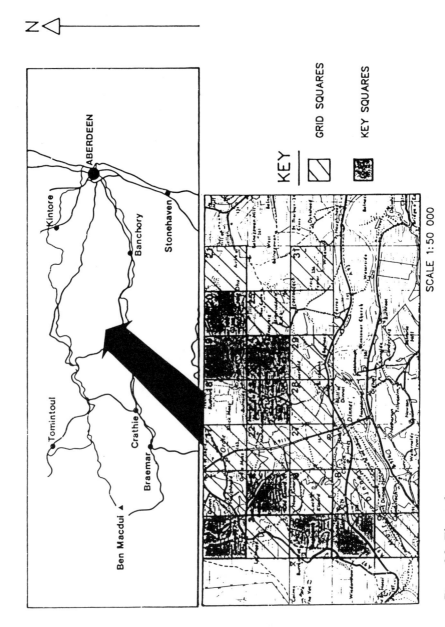

Figure 9.2. The study area.

However, the two levels of coding were complementary and used as one joint code in analyses. For example, two polygons could have identical scores in the 'Percentage cover' columns but, because of species differences, they might receive different scorings in the 'Code' column, e.g. to distinguish bogs ('Code' column, B) from flush grasslands (F), or plantation pines (T) from naturally regenerated pines (3).

Lower-case letters were ascribed to woody vegetation types to describe the dominant growth phase present; for dwarf shrubs these followed the criteria of Watt (1955). Although the interpretation of these codes was very subjective, they proved to be useful indicators when assessing some grasslands which may have been a transient stage after heather burning, and also when examining the older photos for the presence of young (pioneer) woodland. The codes were not used in any analyses.

The 'mosaic' descriptor (V) was used for polygons containing two or more vegetation types in discrete patches where the patch sizes were smaller than the minimum mappable unit of 10 m × 10 m. This information was important because the ecological properties of mosaics of vegetation differ from those of intimate mixtures of species and this could affect the rates and pathways of successions.

Aerial photo-interpretation methodology

A schematic summary is shown in Figure 9.4. Briefly summarized the procedure for each 1 km² grid square was as follows:

1. preliminary stereo-viewing to facilitate familiarity with the vegetation types and the nature of the terrain in the square.

Table 9.1. The land-use/vegetation classification scheme used in the study.

Code	type of vegetation/land use
A	Arable
B	Bog
C	Cytisus (broom)
D	Dry heath
E	Enclosed/improved grassland
F	Flush/wet grassland
G	Unimproved grassland
J	Juniper
K	Rock
L/SW	Loch/standing water
M	Mineral
O	Organic
P	Pteridium (bracken)
R	Rushes
S	Scrub (broom/gorse)
T	Plantation (coniferous)
U	Ulex (gorse)
V	Variable (mosaic)
W	Wet heath
1	Birch
2	Rowan
3	Pine
4	Mixed
5	Coniferous
6	Deciduous
7	Sallow

ITE PROJECT (934)

(Geog Dept	D = dry heath	T = plantation	1 birch
Aberdeen Univ)	W = wet heath	U = ulex	2 rowan
	P = pteridium	C = cytisus (broom)	3 pine
DINNET	G = unimproved grass	O = organic	4 mixed
─────	R = rushes	M = mineral	5 coniferous
Square No.	B = bog	K = rock	6 deciduous
GR	A = arable	V = variable	7 sallow
Year	E = enclosed/impvd	L/SW = loch/	
	grass	standing water	
	S = scrub (broom/		
	gorse)	p = pioneer	
....	J = juniper	b = building	
		m = mature	GT
		d = degenerate	PI

PATCH NO.	CODE	> 50	25–50	10–25	< 10	MOSAIC	SPECIES/NOTES

Figure 9.3. A sample data-coding sheet.

2. mapping of the vegetation polygons, using a 0.1 mm pen nib, onto clear plastic sheet overlaid directly onto the photographs under a stereoscope. Only those polygons that were ≥ 10 m in width/length (estimated by eye) were mapped. Each polygon was then numbered on a photocopy of the overlay (Figure 9.5).

3. classification of each vegetation polygon identified in (2) using the scheme outlined in vegetation classification.

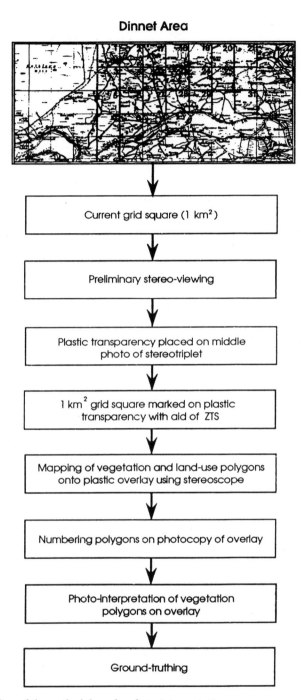

Figure 9.4. Outline of the methodology for photo-interpretation

4. transfer of the overlay to a new transparency to match the Ordnance Survey (O.S.) 1 : 10 000 maps using a Bausch and Lomb Zoom Transfer Scope (ZTS).

5. the twenty-four (i.e. 8 squares × 3 years) 1 : 10 000 overlays were then re-drafted to a scale of 1 : 7500, using an Artograph enlarger, onto 2 sheets of

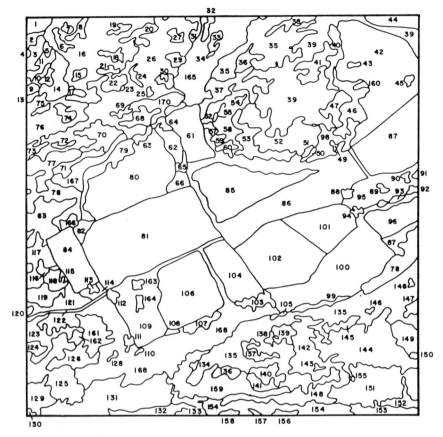

Figure 9.5. Polygon numbering.

AO size drafting film ready for submission to a scanning digitizer. The main reason for the enlargement was that it permitted a *pro rata* increase in the width of pen used for drawing the boundaries. Previous experience had shown that the scanner did not consistently detect a 1.1 mm wide line, resulting in broken lines which caused a considerable increase in subsequent editing. Compositing all the overlays onto two sheets greatly reduced the cost of the digital scanning.

Ground-truthing of the vegetation interpretation

With only eight 1-km² grid squares selected for the final study, it was possible to examine nearly all the polygons mapped on the overlays of the 1985 photographs by walking over the area. Only those polygons that were inaccessible were assessed from a distance (using binoculars) or by comparisons with other polygons deemed to be similar from the photo-interpretation.

Transfer of the polygon boundaries and vegetation classification data to computer files

The methodology that follows was developed under two constraints:

1. the difficulty of and the time involved in accurately hand-digitizing large numbers of very small polygons with complex boundaries; and

2. the absence locally of a suitable geographic information system (GIS). However, large amounts of time were available on a mainframe computer via the Joint Academic Network (JANET) links.

Polygon boundaries

Two steps were involved here. In the first step a flat-bed scanning digitizer was used to convert the polygon boundary drawings on the two sheets of drafting film to raster format digital data stored on magnetic tape. These data were then downloaded onto a SysScan integrated computer system for map information management where they were converted to vector format for editing using the graphic interactive information system (GINIS) function. This function permits the generation, editing, storage and display of digitized polygon files, and was the most efficient way of 'editing-out' and 'editing-in' any extraneous/missing boundaries that had arisen during the scanning process or that differed from the original hand-drawn overlays. Each scanned sheet of 12 overlays was digitized as a single computer file; the 12 individual maps were extracted using other SysScan functions and stored as single files.

The second step was to convert the edited polygon files back to raster format with a pixel size equivalent to 1 m^2 and to transfer them via an IIS system to a mainframe computer. All further manipulation of the map data, including any minor editing not carried out in GINIS, was carried out on the mainframe using a suite of specially written Basic and Fortran programs.

Vegetation classification data

The coded vegetation classifications corresponding to each polygon and grid square were entered manually into a series of Basic data files on a BBC Master Series microcomputer. The individual files were then 'uploaded' onto the mainframe using KERMIT (file transfer software).

Computer processing of the polygon and vegetation files

Polygon files

At this point, all the pixels within polygons in the square had the same code (ϕ). The next step was to recode all the component pixels of each polygon with a single letter and obtain a hard copy so that the individual polygons could be cross-referenced to their original vegetation codings. This section of the procedure, shown schematically in Figure 9.6, involved the following steps.

(1) Conversion of the data from IIS format to ASCII and, finally, binary format (boundary pixels, 1; pixels within boundaries, 0). This was necessary because the IIS output was unsuitable for the Basic and Fortran programs to be used subsequently.

(2) Resizing the polygon files in each set to the same row and column dimensions (a 'set' comprised the files of the 3 years for each square). This was necessary because the processing so far had resulted in small variations (<10 pixels) in the dimensions of files within each set. To carry out the pixel-by-pixel transition determinations later, it was necessary that the files were dimensionally identical.

(3) Checking the files for errors and editing if necessary.

(4) Using a Fortran program to assign an arbitrary number to each polygon and to all the pixels therein. This number would be cross-referenced later to the polygon number assigned manually during the mapping procedure. Although this is

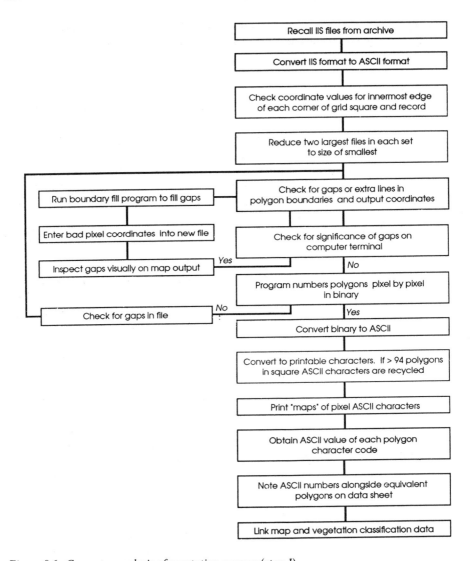

Figure 9.6. Computer analysis of vegetation squares (step 1).

apparently a rather convoluted process, it was the quickest method available for linking the polygons with their original vegetation codings.

(5) Checking the number of polygons in the computer files with the number mapped on the overlays.

(6) To obtain hard copy, the ascribed (binary) pixel numbers were converted to ASCII codes. Each pixel was then assigned the printable character equivalent to the ASCII code; when the number of polygons exceeded 94 (the maximum number of printable ASCII characters), the characters were recycled.

(7) Output the files of printable characters, which form polygon maps, via a line printer.

(8) Numbering the polygons on the map output using the ASCII values of the printed characters. If the character was being recycled, 94 was added to the polygon number.

(9) Matching the polygon numbers from the hand-drawn overlay to the computer-generated polygon numbers and recording on original data sheets.

Vegetation files

The steps involved here are briefly described below and are summarized in Figure 9.7.

(1) Checking and editing the vegetation files and editing in the appropriate computer-assigned polygon numbers determined in (9) above.

(2) Removal of the lower-case letters (growth-phase information) from the classification. The program used for this also checked the files for errors such as incomplete data lines, non-valid vegetation codes, non-printing errors and format errors that could not be trapped elsewhere.

(3) Sorting the vegetation files by ascending computer-assigned polygon number.

(4) Next, the option was available to condense the data into 5 m × 5 m or 10 m × 10 m pixels. The present analyses were done on 5 m × 5 m pixels which ensured that even a polygon of the minimum mappable size of 10 m × 10 m was represented by at least one pixel in the condensed data set. The condensed pixels

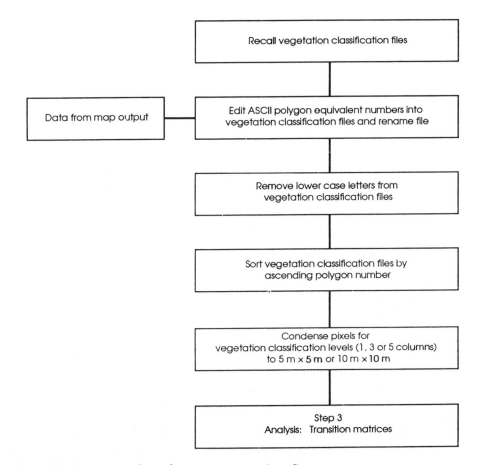

Figure 9.7. Computer analysis of vegetation squares (step 2).

were given the code of the predominant vegetation type of their component pixels, excluding boundary pixels. Where there were equal numbers of two or more different vegetation types, the condensed pixel was coded to the first co-dominant type encountered amongst its components.

(5) The level of the vegetation classification to be used could also be selected and currently we have considered only those vegetation types covering more than 25% of the ground within a pixel (this is the highest level of classification which copes with co-dominant vegetation types). Thus any two polygons would be treated as having distinct vegetation types if their codings differed by one or more letters in the 'Code', >50% or 26–50% columns on the data sheets. After the data had been extracted down to the level required from the three files (years) of each square, a list of vegetation codes without duplication was extracted for each pair of years to be compared in a transition matrix.

(6) Comparing equivalent pixel codings for each of the years 1947, 1964 and 1985 to construct the three transition matrices for each grid square.

(7) The analysis of results included several simple statistical analyses (for example, of changes in areal cover, and rates and frequency of change along different pathways) carried out by straightforward manipulation of the transition matrices produced thus far. Further analyses, which have yet to be undertaken, include examining how successions are affected by the distance to different seed sources, and by spatial and vegetational heterogeneity.

Results

It is not within the scope of this chapter to examine and discuss fully the results obtained so far. For the purposes of illustration only, a highly simplified example of the results from just one 1 km² is presented. The vegetation classification has been reduced to a single dominant vegetation type and the six most common types involved with changes to birch and pine woodland have been selected. Care must be taken in interpreting these simplified results since the pixels are classified only by the dominant vegetation type within a polygon. For example, a polygon with 55% of its pixels in vegetation type A and 45% of type B would be classified as A. However, if the vegetation changes, such that the proportions are reversed, and the polygon is classed as B, it does not imply that all pixels in type A in the polygon have changed to type B; in fact only 10% have changed. It should also be noted that the category 'grassy heaths' included recently burned moorland; whether this reverts to heathland depends very much on local grazing pressures. The 'heath' class contains both wet and dry heaths.

More detailed estimates of changes in the various vegetation components are obtained by analysing the data using several levels of the vegetation classification.

Maps of the areal cover of birch-, pine- and heath-dominated vegetation for each of the 3 years are shown in Figures 9.8 to 9.10, along with their areas as a percentage of the whole square (Figure 9.11). There have been marked increases over the years for birch (from 9% of the square in 1947 to 27% in 1985), unimproved grassland (3% to 18%) and bog (6% to 14%, after dropping to 3% in 1964). Although pine trees were present in the square in 1947, they were not dominant anywhere, and therefore show as zero cover at this level of classification. By 1985, pine trees had become dominant in 5% of the square's area. In contrast, the amount

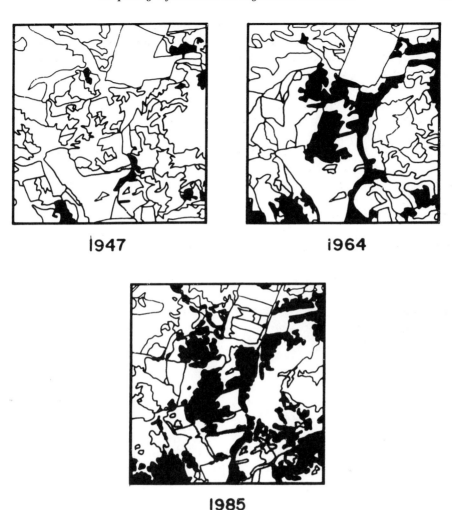

1947   1964

1985

Figure 9.8. Grid square 23—birch (1947–1985).

of moorland had declined from 46% to 11%; only 2% of this loss was accounted for by changes to other vegetation types not shown such as arable/improved grassland, forestry plantation, tall shrubs and bracken. Only in the case of birch did the rates of change approach linearity.

Figure 9.12 shows some of the different pathways involved in succession to birch and pine woodland from other major types; the numbers show the percentage of the vegetation type at the foot of the arrow that has changed to the vegetation at the head of the arrow between the 2 years.

The changes from other components of the square to birch-dominated vegetation indicates the invasive nature of this tree and its ability to establish in widely differing vegetation types. Also of interest are the changes from moorland to only three other types in 1947–1964, but to five other types, including pine dominant, in 1964–1985. These changes cannot be discussed fully here, but some can be related to decreases in grazing pressures and/or the amount of burning taking place (e.g. pine establishing on moorland) and to local increases in grazing pressure (e.g. dry heath

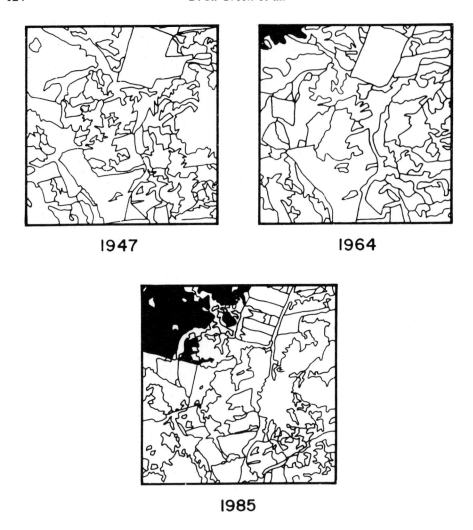

1947   1964

1985

Figure 9.9. Grid square 23—Scots pine (1947–1985).

changing to unimproved grassland, wet heaths to boggy graminoid-dominant vegetation). The aerial photographs showed that some areas were drained before 1964, which partly explains the successions from bogs in Figure 9.12(b).

Fuller discussion of the forces driving these sorts of changes can be found in the literature, e.g. Miles (1988) and Ball *et al.* (1982).

Discussion

Some points arise from the methodological aspect of this work. Firstly, checks between photo-interpretation in the laboratory and subsequent ground-truthing showed that the majority of the vegetation types of interest defined in the classification scheme can be recognized on the photographs selected. For example, dry heath, bracken (in large patches) and grass can consistently be identified, as can

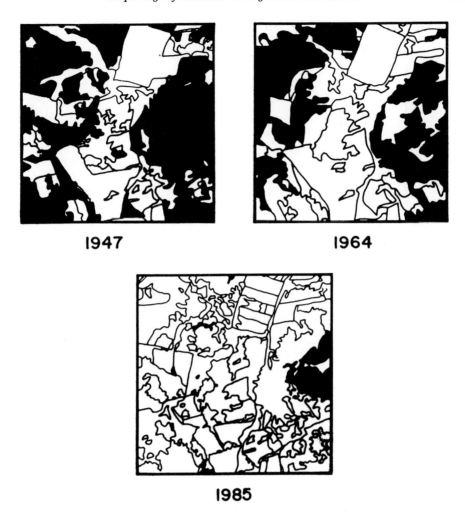

1947   **1964**

1985

Figure 9.10. Grid square 23—dry heath (1947–1985).

birch and pine trees above about 2 m in height. However, even experienced obser-vers have difficulty consistently identifying small patches of scrub vegetation. Gorse, broom, juniper and willow were found to occur in patches that were too small to permit conclusive identification all of the time. Where small stands of scrub have apparently persisted to the present day, it is sometimes possible to confirm that the species has remained the same by ageing the present plants from ring counts of stem sections or cores and seeing if their age predates the earlier photograph. This check is less useful for gorse and broom which rarely live for more than 20 years.

Secondly, although a 'working' methodology has been developed for studying vegetation successions, a number of improvements could be made. Aerial photo-graphy clearly contains the ecological information of interest but the aerial photo-interpretation can be difficult and time consuming; ideally this would be automated. Unfortunately, automated analysis of remotely sensed data (and particularly aerial photography) is not yet at a stage where it can be used satisfactorily for this type of work. Investigations carried out within the framework of this project into alterna-tive sources of data from satellite and airborne scanner systems have shown that

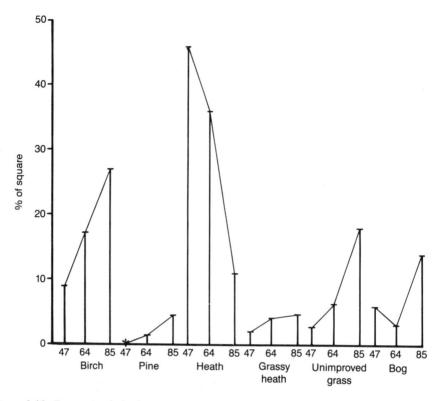

Figure 9.11. Proportion (%) of square occupied by major semi-natural vegetation types in 1947, 1964 and 1985.

they do not yet provide adequate resolution for the purpose of modelling vegetation successions. Furthermore, vegetation classification based solely on spectral information is not sufficiently well developed even today for the classes of most interest. Our conclusions support those of Budd (1987). Improvements are possible with the use of additional information, e.g. texture, which is already incorporated in some automated classification systems, but other factors used during visual photo-interpretation (such as association, shape and size) are more difficult to incorporate.

At present, therefore, even with the currently available image-processing algorithms, satellite data are at best only a supplementary/complementary source of information to standard aerial photo-interpretation. Obviously it will be a long time before we have 40 years of reliable, high-resolution satellite data comparable to the archival aerial photography now available.

Useful improvements can readily be identified in the data-processing aspects of our methodology, particularly with regard to the polygon maps (Figures 9.8 to 9.10); obviously some of the editing procedures could have been conducted more efficiently if we had had continual access to a suitable image analysis system. However, the procedure which took most man-hours (as opposed to computer time) was in linking the unnumbered polygons in the computer files with their appropriate numbers, and hence vegetation classifications, ascribed during the manual mapping. Ideally, one would display a map on a monitor, use a cursor to select any point within a polygon and then input the original number of the polygon whereupon all the component pixels would receive that coding. The ability to do this

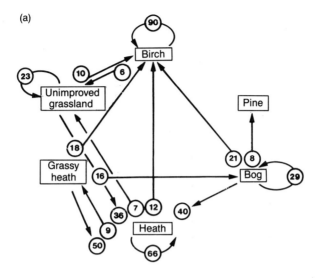

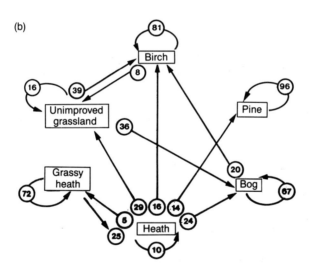

Figure 9.12. *Some of the different pathways involved in succession to birch and pine woodland from other major types.* (a) 1947–1964. (b) 1964–1985.

would delete all the steps in Figure 9.6 from 'Convert binary to ASCII' to 'Note ASCII numbers . . .' and also the steps 'Edit ASCII polygon . . .' and 'Sort vegetation . . .' in Figure 9.7. Our limited investigations suggest that this type of function is not available even on some sophisticated image analysis systems, the classifications either being done during manual digitizing or else the data are received in an already classified form.

In conclusion, it is clear from the preliminary results obtained by this research that, despite constraints imposed by varying scales and quality, it is both possible and practical to use archival aerial photographs to study vegetation successions in upland habitats. The methodology we have developed is functional, although it could be refined, and could potentially be applied to other aspects of remote sensing research.

Acknowledgements

We would like to thank Don French (ITE, Banchory) for writing most of the programs that were essential for this study, Ruth Weaver (Polytechnic Southwest) for her work and advice in conducting the first half of the work described here, and Nigel Brown (ITE, Monkswood) for his guidance and preparation of data on the IIS and SYSSCAN systems.

References

Ball, D. F., Dale, J., Sheail, J. and Heal, O. W., 1982, *Vegetation Change in Upland Landscapes*, Cambridge: Institute of Terrestrial Ecology.

Budd, J., 1987, Remote sensing applied to the work of the NCC in upland areas, in *The Ecology and Management of Upland Habitats: The Role of Remote Sensing (Remote Sensing Special Publication No. 2)*, Aberdeen: Department of Geography, University of Aberdeen.

Fuller, R. M., 1981, Aerial photographs as records of changing vegetation patterns, in Fuller, R. M. (Ed.), *Ecological Mapping from Ground, Air and Space (ITE Symposium No. 10)*, pp. 57–68, Abbots Ripton, Huntingdon: Institute of Terrestrial Ecology.

Hume, E., McMorrow, J. and Southey, J., 1986, Mapping semi-natural grassland communities from panchromatic aerial photographs and digital images at SPOT wavelengths, *International Archives of Photogrammetry and Remote Sensing*, **26**(4), 386.

Jewell, N. and Brown, R., 1987, The use of Landsat thematic mapper data for vegetation mapping in the North York Moors National Park, in *Ecology and Management of Upland Habitats: The Role of Remote Sensing (Remote Sensing Special Publication No. 2)*, Aberdeen: Department of Geography, University of Aberdeen.

Marren, P., 1980, *Muir of Dinnet: Portrait of a National Nature Reserve*, Aberdeen: Nature Conservancy Council.

McMorrow, J. and Hume, E., 1986, Problems of applying multispectral classification to upland vegetation, *International Archives of Photogrammetry and Remote Sensing*, **26**(4), 610.

Miles, J., 1982, Vegetation and soils in the uplands, in Usher, M. B. and Thompson, D. B. A. (Eds), *Ecological Change in the Uplands (Special Publication Series of the British Ecological Society, No. 7)*, pp. 57–70, London: Blackwell.

Morton, A. J., 1986, Moorland plant community recognition using Landsat MSS data, *Remote Sensing of Environment*, **20**, 291.

Watt, A. S., 1955, Bracken versus heather. A study in plant sociology, *Journal of Ecology*, **43**, 490–506.

Weaver, R. E., 1987, Using multispectral scanner data to study vegetation succession in upland Scotland, in *Ecology and Management of Upland Habitats: The Role of Remote Sensing (Remote Sensing Special Publication No. 2)*, Aberdeen: Department of Geography, University of Aberdeen.

Wyatt, B. K., 1984, The use of remote sensing for monitoring change in agriculture in the uplands and lowlands, in Jenkins, D. (Ed.), *Agriculture and the Environment (ITE Symposium No. 13)*, Abbots Ripton, Huntingdon: Institute of Terrestrial Ecology.

10

Landscape structural analysis of central Belgium using SPOT data

H. Gulinck, O. Walpot and P. Janssens

Introduction

The object of landscape research is the fabric of the visible outdoor environment and its role in natural and human-related processes. Landscape planning tries to improve or safeguard these processes through interference in the relevant structures and patterns of the environment. Depending on the viewpoints and ambitions of the professionals, emphasis is put on aesthetic, ecological, environmental, economic or integrated arguments.

The rationale behind landscape research is that such processes, be they of physical, ecological, psychological or technical kind, are to a certain degree controlled by the spatial organization of their environmental setting. This organization is called 'landscape'.

The success of any information system for landscape research is to be judged on the basis of its capability to distinguish or model the elements, components and structures or patterns of the landscape, depending on a specific point of interest.

The great diversity in landscape research can be appreciated not only through the above-mentioned professional arguments, but also through the different categories of fabric, structure or pattern in the landscape:

1. geographical units (ecotopes, soil units, landscape facets, etc.);
2. structures of linear elements or processes (roads, field boundaries, orientations, material flows, etc.);
3. spatial heterogeneities (land-use complexes, patterns in natural vegetation, etc.);
4. visual structures (viewsheds, visual patterns, textures and colours, landscape mystery, etc.);
5. functional organizations (farming landscapes, dwellings, infrastructures, etc.); and
6. objects and patterns of social value (historic landmarks, recreation patterns, etc.).

The need for information about the landscape on detailed levels of spatial definition, i.e. low levels of aggregation, is common. Landscape research and planning is very information intensive, especially in the spatial dimension. Unfortunately, landscape details are often available or accessible within limited sites only. Landscape data are, however, considered increasingly as important environmental arguments at regional and higher levels as well, but very often information is only available in aggregated form and quality.

In this chapter, an approach to the investigation of landscape structures at a regional level is illustrated. Classified remote-sensing data are used as input for the application of a series of image analysis algorithms, used for the modelling of functional landscape patterns.

For the implementation of remote-sensing data in landscape research we developed a theoretical framework, with two main stages, the analytical stage and the synthetical stage. In the analytical stage, relevant elementary landscape data (land-use classes, and landscape objects) are derived from SPOT data or other spaceborne images through multispectral/multitemporal classification procedures and image object recognition algorithms. Multispectral SPOT data have a resolution of 20 m × 20 m and cover areas of 60 km × 60 km. In the synthetical stage the landscape information elements from the analytical stage are synthesized to models of landscape patterns and structures.

In this chapter, the organizing principle will be illustrated for the investigation of landscape ecological structures.

Detection and reconstruction of the ecological relevant land-use categories

The study area (20 km × 60 km) is located in central Belgrium (Figure 10.1). SPOT data (SPOT XS KJ 43/247 16-08-1987) were classified with the maximum likelihood procedure. The classification was compared with an updated ground truth map of the study area and further iteratively improved. The seven classes retained (two urban classes, crops and grassland, bare fields, water, deciduous trees, and coniferous trees) were regrouped according to the study objectives. We considered all woody vegetation as relevant ecological categories.

This method yielded up to 95% correct mapping of existing woodland. For woodland strips the minimum diameter for identification is 10 m, shrub patches and strips are identified as wood if the minimum width exceeds 15 m. Some linear features smaller than 20 m but taller than 15 m (especially poplar trees) could be

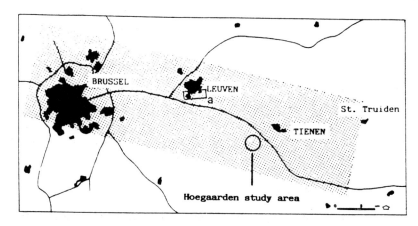

Figure 10.1. Study area in central Belgium.

classified in relation to their orientation. North-east orientation, perpendicular to the sun-azimuth at the moment of imaging, yields better classification. The maximum likelihood classification results, especially for the marginally sized elements of land-use classes, depend mainly on the selection of spectral channels, image dates, training sampling and threshold levels. For a first approach in landscape structural analysis, where we are mainly concerned with contrasts such as mass-space, high-low vegetation, built-open, the seven retained classes are suitable.

The detection of vegetated linear elements (hedgerows, sunken roads, streams, etc.) is a particular issue in the classification of remote-sensing images. Such linear elements are important in several landscape functions such as visual perceptions, surface morphodynamics, habitats, refuges and corridors for biota. Because of their narrow width in general, in SPOT data linear elements often appear as mixed pixels, characterized as mixed signatures generated by two or more land-use units. Using the image resolution as scale reference, mixed pixels could be considered as indicative for 'microheterogeneity' of the landscape. In the case of multispectral SPOT this is a scale range of 10–30 m. In a former paper (Gulinck *et al.*, 1988), methods and results for the mapping of small wooded linear elements are described.

Structural interpretation of a classified image

Satellite images are unique in the sense that they provide recent landscape information for a large area in raster format with a relatively high accuracy. There are many landscape researchers and ecologists who work with land-use data in grid format for the derivation of landscape patterns (Turner and Ruscher, 1988; O'Neill *et al.*, 1988; Milne *et al.*, 1989). Spatial characteristics such as proximity, connectedness, patch size and form can easily be deduced from such data. The dimensions (measurements) as well as the topographical structuring can provide useful information for the understanding of landscape functioning. In order to structure the ecological survey we distinguish three topics: objects in the landscape, the landscape matrix (open areas in the landscape), and the transition areas between object and matrix.

Different methods can be applied for the development of spatial hypotheses of ecological relevant landscape structures. A few general interpretation concepts were selected on the basis of the similarity between structural principles in image interpretation on the one hand, and landscape ecology on the other. Figure 10.2 illustrates these spatial criteria, their graphical interpretation and the corresponding descriptive landscape variables. The programs developed were written in VAX/Pascal and were executed on the Minivax 8350.

Objects in the landscape

Using SPOT data, landscape elements or patches (ecotopes) can be detected and their spatial characteristics can be described with the implementation of techniques based on the analysis of contiguity.

Object recognition, edge detection

The application of object recognition identifies each ecological entity within one land-use category. The algorithms developed to calculate landscape relevant indices

IMAGE INTERPRETATION

STRUCTURES IN LANDSCAPE
ECOLOGY

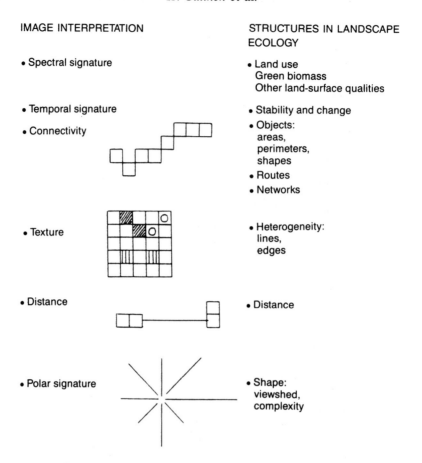

- Spectral signature

- Land use
  Green biomass
  Other land-surface qualities

- Temporal signature

- Stability and change

- Connectivity

- Objects:
  areas,
  perimeters,
  shapes
- Routes
- Networks

- Texture

- Heterogeneity:
  lines,
  edges

- Distance

- Distance

- Polar signature

- Shape:
  viewshed,
  complexity

Figure 10.2. Methodological principles.

such as patch size, patch shape, patch perimeter, number of patches, isolation and connectivity (Forman and Godron, 1986) are based on the distinction between edge pixels and object pixels. Figure 10.3 illustrates the different frequency distributions derived for each variable in the study area shown in Figure 10.1. The shape index (V) is measured using the formula:

$$V = \frac{4\pi A 100}{P^2}$$

where A is the area, P is the perimeter and $\pi = 3.14\,519$. The shape index is 100 for a circle.

Object clumping

Distance is an important parameter in the study of mobility of organisms within corridors and between isolated landscape elements (Van Dorp and Opdam, 1987; Burel, 1989). To measure the distances between objects a radial search-line procedure in different directions from the woodland edge pixels to the other woodland objects is developed. Depending on the critical distance chosen according to the

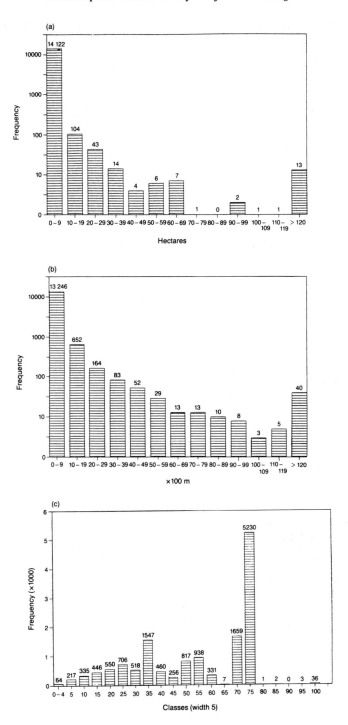

Figure 10.3. Frequency distributions calculated for woodland areas in the study area shown in Figure 10.1. (a) Area; (b) perimeter; (c) shape factor.

Figure 10.4. Patterns of woodland patches less than 20 m isolated from other woodland patches in the Hoegaarden study area. Note the two main clusters.

mobility ranges of species or individuals, different structures of ecological continuity can be visualized. Figure 10.4 shows an analysis of grouped woody areas within a distance of 200 m in the Hoegaarden test area. The analysis was done with 16 search directions out of each border pixel.

Object orientation

The geographical orientation of landscape elements is another ecological variable. In practice this variable is ecologically relevant in combination with other topographic characteristics (e.g. the orientation in relation to hill exposure). Orientation is measured as the direction of the largest length of the object.

Landscape matrix

The study of the landscape matrix is ecologically complementary to the study of the objects. This matrix, often the open agricultural area between semi-natural ecotopes or woods, has an influence on the overall ecological processes (migration, etc.). In order to characterize this 'open space', we used the technique based on a radial searchline procedure applied in each pixel preclassified as 'open space'. This technique is developed to describe the landscape matrix in terms of two spatial descriptive and statistical independent parameters size and shape. The mathematical principle is illustrated in Figure 10.5.

There is a tendency in landscape ecological research to refer to theories of human perception and design, and vice versa (Bartkowski, 1984; Nassauer, 1988).

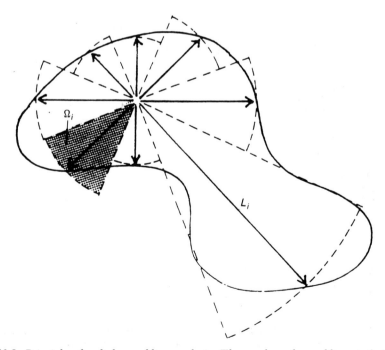

Figure 10.5. Principle of radial searchline analysis. The number of searchlines is 8, L_i is the length of the measured distance in direction i, Ω_i is the area of the section of the search circle defined by the length in direction i.

Intervisibility is in this sense an important characteristic in the assessment of landscape qualities. Intervisibility is related to distance, but includes blocking elements such as topography and high raising land use.

Our interest is to compare the structural distinction derived from an environmental analysis as observed from above (the techniques previously explained) with the perceptual characteristics in the environment obtained from within the landscape. This implies the completion of SPOT data with a digital elevation model (DTM). Taking into account the logarithmic characteristic of visual perception and linking an elevation model to the classified SPOT data we can simulate the visual order and complexity (variety) in the landscape. The measurement of open spaces is visualized with a combination of a shape index and the mean distance measured in *n* directions. The shape index is calculated as follows:

$$\text{Shape index} = \frac{\Sigma L_i}{R_{eq}}$$

where L_i is the length of the distance in the direction i, and R_{eq} is the radius of the circle with an area equal to the measured area.

This technique also enables one to map the visual and geographical urbanization of the landscape or the proximity from a certain point in the landscape to the elements of any land-use class. Figure 10.6 shows the visibility of woodland areas in a test area of Hoegaarden.

Transitional zones

Texture analysis

Texture classification was used in the first method allowing a rapid appraisal of the composition of landscape objects and their surrounding matrix. Texture classification covers a wide range of methods and techniques, differing in the definition of texture itself and its algorithms. A simple method used here consists of a frequency analysis of different categories present in windows of 3×3, 5×5 or larger pixel size. The frequency of the different land uses in the window determines the classification of the central pixel. This texture analysis enables one to classify compositional heterogeneities and transitional zones (Rogala, 1981). The size of the window, weighting factors and frequency profiles are important parameters. The window size should be selected in relation to image resoluion and ecological size considered. Figure 10.7 illustrates the geographical woodland compositions calculated within 5×5 pixel windows for the Hoegaarden test area (Figure 10.1). Seven equal classes from 0% to 100% woodland occupation were distinguished.

Conclusion

The results derived from the procedures described can be considered as new databases which enable the classification and assessment of landscape structural data. It encourages the cartographic representation of structural landscape data which can be considered as a working basis for testing landscape ecological hypotheses. Because of the systematic synthesis of quantitative landscape measurements the results can assist in landscape planning and design processes.

0   **500**   **1000 m**

1   2   3

Figure 10.6. Visibility range of woodland in the Hoegaarden study area calculated on the combination of a classification and a digital relief model. (1) Visibility area of woodland; (2) woodland; (3) woodland not visible.

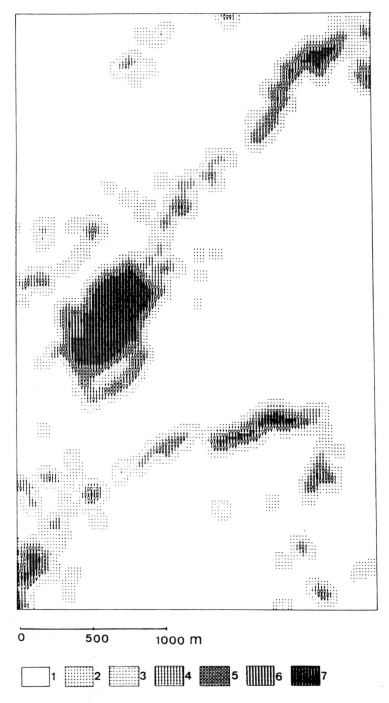

Figure 10.7. *Texture analysis for woodland presence in the Hoegaarden study area within 5 × 5 pixel windows (100 m × 100 m), centred over each pixel. (1) 0% woodland; (2–7) 1–100% equidistant classes.*

Note: this paper presents results of the Belgium Scientific Research Program in the field of spatial remote sensing (Services of the Prime Minister—Science Policy Office). The scientific responsibility is assumed by the authors.

References

Bartowski, T., 1984, *The Relation Between Landscape Ecology and Landscape Perception, IALE Proc. First International Seminar on Methodology in Landscape Ecological Research and Planning, Theme 1*, pp. 113–117, Roskilde: Roskilde Universitetsvorlag GeoRuc.

Burel, F., 1989, Landscape structure effects on carabid beetles spatial patterns in western France, *Landscape Ecology*, **2**, 215–225.

Forman, R. T. T. and Godron, M., 1986, *Landscape Ecology*, Chichester: Wiley.

Gulinck, H., Janssens, P. and Walpot, O., 1988, *Teledetectie in Landschapsonderzoek en -Planning*, Symposium Wetenschappelijk Onderzoeksprogramma inzake Teledetectie per Satelliet, Brussels: 5–6 December (in press), Diensten von de Erste Ministe.

Nassauer, J. I., 1988, *Perceptual and Ecological Concepts of Landscape Corridor*, Minnesota: University of Minnesota.

Milne, B. T., Johnston, K. M. and Forman, R. T. T., 1989, Scale-dependent proximity of wildlife habitat in a spatially-neutral Bayesian model, *Landscape Ecology*, **2**, 101–110.

O'Neill, R. V., Krummel, J. R., Gardner, R. H., Sugihara, G., Jackson, B., DeAngelis, D. L., Milne, B. T., Turner, M. G., Zygmunt, B., Christensen, S. W., Dale, V. H. and Graham, R. L., 1980, Indices of landscape pattern, *Landscape Ecology*, **1**, 153–162.

Rogala, J.-P., 1981, *De l'Image vers les Cartes Thématiques, 4ème Colloque International du G.D.T.A.*, pp. 56–66. France: Groupment Pour le Développement de la Télédétection Aerospatiale.

Turner, M. G. and Ruscher, C. L., 1988, Changes in landscape patterns in Georgia, USA, *Landscape Ecology*, **4**, 241–251.

van Dorp, D. and Opdam, P. F. M., 1987, Effects of patch size, isolation and regional abundance on forest bird communities, *Landscape Ecology*, **1**, 59–73.

11

Using cover-type likelihoods and typicalities in a geographic information system data structure to map gradually changing environments

T. F. Wood and G. M. Foody

Introduction

The incorporation of vegetation classifications derived from the analysis of remotely sensed data into geographic information system (GIS) structures has become common practice. However, the usefulness of classifications derived from satellite remotely sensed data as an input may be questioned for those applications concerned with environments that display gradual change (Jupp and Mayo, 1982; Frank, 1984; Allum and Dreisinger, 1987; Griffiths *et al.*, 1988). This is because classification results in elements being 'true' or 'false' for a class, whereas their membership is frequently *uncertain* between classes. Furthermore, the derivation of land-cover classes from the analysis of satellite remotely sensed data by frequently used maximum likelihood methods does not provide all of the information generated during the classification process (Trodd *et al.*, 1989). This is because relative likelihoods (posterior probabilities of class membership) are replaced by the code of the class with the highest ranked likelihood as the final output of the classification.

Likelihoods derived from remotely sensed data indicate the relative strength of class membership across the classes being considered (Thomas *et al.*, 1987) and can thus be used to map the variations in class likelihood spatially (Wallace and Campbell, 1988; Wood and Foody, 1989). The 'typicality' of each element refers to the probability of its spectral characteristics with reference to each class parent population in turn (Campbell, 1984a) and this is also obtained during the classification process. Typicality and likelihood may be used jointly in image display of information derived from remotely sensed data (Wallace and Campbell, 1988). Furthermore, they may be incorporated usefully into a GIS data structure to provide information on the certainty of allocation between classes, within-class typicalities, and gradients of change between classes. Ancillary data and contextual information also held within the GIS may be employed with class likelihoods to improve allocation accuracy (Di Zenzo *et al.*, 1987).

Errors in mapping uncertain features may be reduced through the use of likelihoods in the GIS in place of most likely classes (the output from conventional image classification procedures). The inclusion of a class code derived from remotely sensed derived data with a set of other attributes for an element which has spectral properties similar to those of more than one class greatly increases the probability of inherent error (Walsh *et al.*, 1987) when applying the GIS rule base to that element. For example, to map 'wet heath' the rule base may employ the class 'wet heath', a substrate category and an altitude category in order to define an area. Levels of confidence in the accuracy of substrate and altitude classes are effectively constant for each element considered, whereas the confidence in the class 'wet heath' will vary with the absolute value of the maximum likelihood used to derive the class, and with the difference between the two highest likelihoods. Furthermore, error will also be introduced from the inclusion of a fixed class boundary for a cover type that would be classed differently at different times of the year. Hence the extent of the 'wet heath' class is better represented as a varying surface within the GIS. The availability of likelihoods enables selection of type I and type II error margins according to user needs.

It is therefore suggested that the completion of a classification before data input into the GIS is premature. An alternative is to build likelihoods and typicalities of cover types derived from the analysis of remotely sensed data into the database together with ancillary information. It is then possible to carry out a sequence of operations on the database within the GIS in which likelihoods indicate relative as opposed to absolute class membership for each element.

Derivation of typicalities and likelihoods

The maximum likelihood classifier derives a probability density function for each class under consideration by using statistics obtained from a training set of sample pixels:

$$p(x\,|\,i) = [p1/2\pi^{n/2}\,|\,V_i|^{1/2}]\,\exp\,[-0.5(D)].\tag{11.1}$$

where $p(x\,|\,i)$ is the probability density function for pixel x with respect to class i, $p1$ is the prior probability, V_i is the variance covariance matrix for class i, n is the number of bands and D is the Mahalanobis distance from x to the centroid of i.

Given a total likelihood of 1.0 for each pixel across all classes (g), likelihoods are derived from the decomposition of total likelihood relative to individual class probability densities (Campbell, 1984a; Thomas *et al.*, 1987)

$$m1(x\,|\,i) = p(x\,|\,i)\bigg/\sum_{i=1}^{g}(x\,|\,j)\tag{11.2}$$

The Bayes optimal strategy is used to obtain the maximum likelihood from the comparison of likelihoods or of probability densities under certain conditions (Mather, 1987; Thomas *et al.*, 1987). In the absence of any weighting of likelihoods through typicality (Campbell, 1984b), class allocation can be based on small class likelihoods (Foody *et al.*, 1988). Typicalities may be derived from referring D values to the χ^2 distribution with n degrees of freedom. A pixel may be atypical of all classes under consideration and thus be excluded from classification.

Mapping variations in the heathland environment

The division between woodland and heathland is sometimes obscure for field mapping purposes due to varying amounts of scrub (Nature Conservancy Council,

1985) and even more so for remotely sensed data. Similarly, moisture gradients in heathland result in complex gradations in community not easily represented by simple categories. Heathland may be considered to display gradual changes in community type both spectrally (Wardley *et al.*, 1987) and ecologically (McIntosh, 1967; Huntley, 1979). Application of choropleth mapping to such environments obscures useful information on ecotonal areas where communities merge.

The study area

The Ash ranges site of special scientific interest (SSSI) and the surrounding area are located across the Chobham Ridges, a cap of tertiary plateau gravels resting on Bagshot Sands. The communities of wet and dry heath that are supported (Harrison, 1970) are considered to be of national significance. Moisture gradients, soils, geology and land use are all important influences on the vegetation. The ridge of plateau gravels supports expanses of dry heath and grass heath where burning is of suitable frequency and intensity. Neglect of these areas enables them to be overtaken by bracken, birch, Scots pine and thence to scrub woodland (Nature Conservancy Council, 1985). The lower lying Bagshot Sands to the east support wet heath and bogs of considerable variety with flushes of higher nutrient status. Pine and birch are rapid colonizers of these wet areas when management is abandoned. The steep slopes which mask the edge of the Chobham Ridges support dry heath communities with dwarf gorse; gulleys may contain wet heath where flushing occurs. Bracken invasion is pronounced upon the slope faces.

The construction of the database

Likelihoods and typicalities were obtained for each 30 m × 30 m pixel cell of the Ash ranges SSSI from an analysis of Landsat TM bands 1–5 and 7. A total of 135 training and 135 testing pixels were sampled for each of five cover types, namely: coniferous woodland (CW), mixed woodland (MW), pioneer and burnt heath (PBH), dry heath (DH), and wet heath and bog (WHB). These were derived from the original communities following the grouping of categories of related cover type which also proved to be spectrally indistinguishable (pioneer heath and burnt heath, woodland/glassland mosaic and deciduous woodland, wet heath and bog) and the omission of bracken, which showed confusion with several classes. The data were then used to classify the testing samples on the basis of maximum likelihood decision rules generated by the training data. The five cover types show varying degrees of overlap for the testing samples (Table 11.1) and indicate the expected extent of confusion in the parent populations upon classification.

In addition, information on slope and altitude were derived for each pixel from an analysis of digitized spot heights obtained from an Ordnance Survey 1 : 25 000 map of the study area. Altitude agrees closely with changes in underlying geology in the area, whilst slope has important controls over the degree of flushing in the heath community. The database thus contained the following information for each cell: maximum likelihood class code, class likelihood and typicality for each class (percent probability), altitude and slope (to the nearest integer).

Analysis of the database for habitat mapping

The distribution of training data was examined to determine the altitudinal range in which the wet heath communities of interest are found. This reflects the occurrence

Table 11.1. Confusion matrix for the five cover-types, the upper values are percentages and the lower values are pixel numbers.

	Predicted				
Actual	CW	MW	PBH	DH	WHB
CW	**97.78**	2.22	0.00	0.00	0.00
	132	3	0	0	0
MW	14.81	**82.22**	0.00	2.22	0.74
	20	111	0	3	1
PBH	0.00	1.48	**86.67**	5.93	5.93
	0	2	117	8	8
DH	0.74	6.67	0.00	**85.93**	6.67
	1	9	0	116	9
WHB	0.00	8.15	0.00	17.04	**74.81**
	0	11	0	23	101

Overall accuracy 85.48%.

of a substrate of Bagshot Sand. Similarly, a maximum slope of 23° was also identified for this class from training data. The range of 53–96 m and slope of <23° were then used with maximum likelihoods of wet and dry heath to exclude pixels with an unsuitable altitude and slope or of a different cover-type likelihood from further consideration. Within the areas identified, typicalities of wet heath and dry heath were displayed to map variations, using red and blue colour guns. The resulting image indicates areas of heath atypicality, areas where both classes are typified, and areas where either one or the other is represented within the region where heath is the most likely cover (Wood, 1988).

As is evident from Table 11.1, there is overlap between heath categories that make maximum likelihood separation of these categories prone to error; variations *within* the class of 'heath' may be examined through the overlay of typicalities in order to preserve the information present in the overlap regions, whilst the gradient of cover type change may be mapped using likelihoods. Depending on the statistical nature of the data, however, other measures of probability may be more appropriate for mapping continua (Trodd *et al.*, 1989). The simple example above demonstrates that inclusion of both types of probability surface within a GIS framework gives more flexibility than a maximum likelihood classification alone could do, even though these probabilities are derived as a by-product of the classification procedure. It would also be possible to group typicalities into classes of uncertainty such as 'atypical', or likelihoods into classes such as 'unlikely', in order to simplify their use in selection procedures.

Conclusions

It may be more useful to incorporate probabilities (typicalities or likelihoods) into the GIS than just the class code derived from a maximum likelihood classification when the environment in question displays gradual change in class membership. This may both reduce error and allow greater use of the information derived during maximum likelihood classification procedures.

Semi-natural vegetation in particular may be more appropriately mapped through making use of probability gradients which represent the intergrading that occurs in such regions.

Acknowledgements

We would like to thank Dave Talmage for assistance with image processing and Norm Campbell for valuable discussion at IGARSS '88. We also thank Nigel Trodd for his contribution to the development of the concepts underlying this work.

References

Allum, J. A. E. and Dreisinger, B. R., 1987, Remote sensing of vegetation change near Inco's Sudbury mining complex, *International Journal of Remote Sensing*, **8**, 399–416.

Campbell, N. A., 1984a, Some aspects of allocation and discrimination, in Van Vark, G. N. and Howells, W. W. (Eds), *Multivariate Statistical Methods in Physical Anthropology*, p. 177, Dordrecht: Reidel.

Campbell, N. A., 1984b, Mixture models and atypical values, *Mathematical Geology*, **16**(5), 465–477.

Di Zenzo, S., Bernstein, R., Degloria, S. D. and Kolsky, H. G., 1987, Gaussian maximum likelihood and contextual classification algorithms for multigroup classification, *IEEE Transactions on Geoscience and Remote Sensing*, **25**, 805–814.

Foody, G. M., Wood, T. F. and Trodd, N. M., 1988, Classification decision rule modification on the basis of information extracted from training data, *IGARSS '88 Remote Sensing Moving Toward the 21st Century* (Special Publication No. SP-284), p. 513, Paris: European Space Agency.

Frank, T. D., 1984, Assessing change in the superficial character of a semi-arid environment with Landsat residual images, *Photogrammetric Engineering and Remote Sensing*, **50**, 471–480.

Griffiths, G. H., Wooding, M. G., Jewell, N. and Batts, A. J., 1988, *Use of Satellite Data for the Preparation of Land Cover Maps and Statistics*. Progress Report to the Department of the Environment, April 1988, Farnborough: National Remote Sensing Centre.

Harrison, C. M., 1970, Phytosociology of certain English heathland communities, *Journal of Ecology*, **58**, 573.

Huntley, B., 1979, The past and present vegetation of the Caenlochan National Nature Reserve, Scotland. (1) Present vegetation, *New Phytologist*, **83**, 215–283.

Jupp, D. L. B. and Mayo, K. L., 1982, The use of residual images in Landsat image analysis, *Photogrammetric Engineering and Remote Sensing*, **48**, 595–604.

Mather, P. M., 1987, *Computer Processing of Remotely Sensed Images*, New York: Wiley.

McIntosh, R. P., 1967, The continuum concept of vegetation, *Botanical Review*, **33**, 130–187.

Nature Conservancy Council, 1985, Hicks, M. (Ed.), 'A Heathland Survey of Surrey', Lewes, Sussex: Nature Conservancy Council (unpublished report).

Thomas, I. L., Benning, V. M. and Ching, N. P., 1987, *Classification of Remotely Sensed Images*, Bristol: Adam Hilgar.

Trodd, N. M., Foody, G. M. and Wood, T. F., 1989, Maximum likelihood and maximum information: mapping heathland with the aid of probabilities derived from remotely sensed data, *Remote Sensing for Operational Applications*, Nottingham: Remote Sensing Society.

Wallace, J. F. and Campbell, N. A., 1988, 'Statistical Methods for Cover Class Mapping using Remotely Sensed Data', CSIRO Division of Mathematics and Statistics, Australia (unpublished report).

Walsh, S. J., Lightfoot, D. R. and Butler, D. R., 1987, Recognition and assessment of error in geographic information systems, *Photogrammetric Engineering and Remote Sensing*, **53**, 1423–1430.

Wardley, N. W., Milton, E. J. and Hill, C. T., 1987, Remote sensing of structurally complex semi-natural vegetation—an example from heathland, *International Journal of Remote Sensing*, **8**, 31–42.

Wood, T. F. and Foody, G. M., 1989, Analysis and representation of vegetation continua from Landsat thematic mapper data for lowland heaths, *International Journal of Remote Sensing*, **10**, 181–191.

Wood, T. F., 1988, Methods for the analysis and display of ecotones from Landsat thematic mapper data of semi-natural vegetation, *IGARSS '88 Remote Sensing Moving Toward the 21st Century (Special Publication SP-284)*, p. 929, Paris: European Space Agency.

12

The use of remote sensing (SPOT) for the survey of ecological patterns, applied to two different ecosystems in Belgium and Zaire

R. Goossens, T. Ongena, E. D'Haluin and G. Larnoe

Introduction

This study deals with the transformation of raw satellite data into processed images which reveal certain ecological patterns and structures. The latter topics are only treated briefly and in a general way because the resulting maps are intended as a basis for further detailed research by ecologists. The study areas (Belgium and Zaire) were chosen so that it was possible to evaluate the suitability of the applied methodology for the analysis of different environments.

Techniques

Several supervised classification techniques can be used to detect typical landscape ecological patterns. For this study three techniques were tried out: box classification, minimum distance to mean, and maximum likelihood classifiers (Richards, 1984). The maximum likelihood classifier gave the most acceptable results.

Before classification of the image, an image enhancement is usually necessary. This results in an improvement of the image for a particular application. The enhancement technique used in this study, particularly for the detection of patches, was 'contrast stretching'. To detect linear landscape features some 'edge enhancement filters' were used (e.g. Sobel, Lahe and Roberts) and some of our filters (I^2S-manual, 1987).

Classification

Several classification techniques can be used to detect typical landscape ecological patterns. The techniques used are based on the principle of collecting some pixel values, which can be determined as belonging to a certain class. These classifications are known as *supervised classification techniques*. With the spectral information, collected in different test areas (ground truth), some calculations are performed. Based on the results the complete image is classified according to one of the classification procedures.

A first technique applied was the *box* or *parallelepiped* classifier. This is performed by inspecting histograms of the individual spectral components of the training data. The upper and lower significant bounds on the histograms are identified and used to describe the brightness value range of each component. The range in the three bands together describes a three-dimensional box, or parallelepiped. If pixels are located in such a parallelepiped they are labelled as belonging to that class.

A second classification technique used was the *minimum distance to mean* classifier. Training data were only used to determine class means. Classification is performed by placing a pixel in the class of the nearest mean.

The *maximum likelihood* classifier calculates the probability that a pixel belongs to a certain class. Each pixel is assigned to the class for which the probability is the highest (Mather, 1987).

Enhancement

Before classifying an image, some form of *image enhancement* is usually necessary. This results in an improvement of the image for a particular application. The enhancement technique used in this study, particularly for the detection of patches, was *contrast stretching*. In general, a small part of the possible range of digital number (DN) is used for any particular scene. If these pixels are displayed in their raw state the image will have such a low contrast that it is difficult to differentiate between objects having a slightly different DN. To overcome this problem the measured DNs of the pixels are stretched onto a wider display scale. The minimum and the maximum DN of the original image are pulled to the lowest and the highest value of the new display scale, respectively. In this way intermediate values are altered and the contrast between the different features becomes clearer.

To detect linear landscape features some *edge enhancement filters* are used. This technique puts a matrix, in general a 3 × 3 matrix (kernel), over the image. The pixel values in this kernel are multiplied by a factor depending on the filter used and the summation of the new values is put in the centre of the kernel. Afterwards the kernel shifts one position and the procedure starts again.

Imagery and maps used

Imagery

The images used in the study are shown in Table 12.1.

Maps

The following maps were used in the study.

1. Katenda (Zaire)
   (a) Topographical maps of the IGZ (1956) at a scale of 1 : 50 000; and
   (b) localization map of the region of Katenda at a scle of 1 : 100 000 (1980).
2. Kempen (Belgium)
   (a) Topographical maps of the NGI (1971) at a scale of 1 : 10 000, 1 : 25 000 and
       1 : 50 000.

Table 12.1. Imagery used.

Test site	Date	Scene	Sensor
Katenda	2 July 1987	103/361	SPOT XS 1B
Katenda	24 May 1988	103/361	SPOT XS 1B
Kempen	28 July 1986	043/246	SPOT XS 1B

The elaboration of a global ecological information package: two case studies

Zaire (Katenda)

Environmental setting

One of the test sites for this study is located in Bandundu, a region in the south-western part of Zaire (Figure 12.1). This area can be divided in three subregions: the Maï-Ndombe, a woody and humid zone in the north; the Kwango-area, a dry and sandy plateau in the south; finally, as a transition zone, the Kwilu-area (Figure 12.2). The latter subregion is characterized by many wide (mainly north–south orientated) valleys in the north and a dry plateau in the south.

The vegetation of the Kwilu is mainly related to the soil type. Two different vegetation types are found: the forest and the steppe savannah. The former is located in the valleys on the 'terres rouges' (red clayey sandstone), the latter on the plateau (Kalahari sands). Most of the natural vegetation is degraded by various human influences: shifting cultivation, collecting of fuel wood and fire raising. The activity last mentioned occurs especially on the steppe and the savannah.

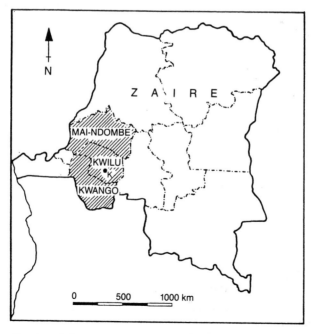

Figure 12.1. Localization of the test site in Zaire. Shaded area—Bandundu region. K, village of Katenda.

Figure 12.2. SPOT-XS band 3 (1988) image of the Kwilu region near the village of Katenda (Zaire).

Detection and mapping

From an ecological point of view a landscape can be defined as 'heterogeneous land area composed of a cluster of interacting ecosystems that is repeated in similar form throughout' (Forman and Godron, 1988: 11).

Consequently, landscape ecology focuses on three characteristics of the landscape:

1. structure—the spatial relationships among the distinctive ecosystems or landscape elements;
2. function—the interactions among the component ecosystems (flows of energy, materials and species);
3. change—the alteration in the structure and function of the ecological mosaic over time.

These three fundamental characteristics are closely related to each other: by studying one of them it is also possible to obtain information and insight concerning the others. Landscapes differ functionally according to their structure. Also, changes in the structure of the landscape are related to changes in its functioning, and vice versa. Finally, the overall structure (heterogeneity) of the landscape influences the biotic diversity, the species flow and the energy flow.

We now focus especially on landscape structure and change. With regard to structure, three basic types of landscape element can be distinguished: *patches, corridors* and *matrix*. According to the kind of element we want to detect, different image-processing techniques are used.

Because the types and dynamics of species and the stability of a landscape element are, to a large extent, determined by the characteristics (origin, size, shape, number, width, connectivity, etc.) of the element itself, the different processed images offer each time a wide range of possible measurements and quantification that can provide relevant ecological information. These possibilities are discussed more in detail in the following section.

Patches

A patch can be defined as 'a non-linear surface area differing in appearance from its surroundings' (Forman and Godron, 1988: 83). Normally, patches in a landscape are plant and animal communities, i.e. assemblages of species. In any area three types of patches can be found: disturbance patches, introduced patches and environmental resource patches.

Disturbance patches are mainly caused by forest and steppe fire. One can distinguish regular and irregular ones. The regular fire patches are used for cattle raising. Fire extension is controlled by fire lanes. The main purpose of this fire raising is to regenerate the grass for the cattle during the rainy season. The irregular patches are uncontrolled or semi-controlled fires. The autochthonous population burns the steppe for hunting purposes. Due to the uncontrolled character of the fires, most of the irregular patches are larger than the regular ones. It is easy to classify these phenomena because of their very strong absorption of the incident light. In some cases, however, water surfaces can erroneously be classified as fire patches. This has not been the case in our region for both periods.

Introduced patches are related to agricultural land. The fields can be recognized due to the fact that the leaf area index (LAI) of the agricultural land is lower than that of the steppe. Using a box classification it is possible to differentiate the fields from the steppe. The class 'fields', however, also include roads, villages and kraals: all are characterized by a very low LAI.

The patches are localized on the plateau and consist of two vegetation types: woodland savannah and open forest. This vegetation can be mapped by a minimum distance classification (Figure 12.3). Four wood vegetation types can be detected: palm trees, gallery forest, woodland savannah and open forest. They must be classified in such a way that the overlap between the different classes is minimal.

The species dynamics and the patch turnover depend largely on the origin of the patch itself. Number, size and shape—which can easily be quantified and measured on processed satellite images—have effects on energy and nutrient storage or flow, on microenvironment and on species composition and abundance ('edge'. species; 'interior', species) by the edge effect (Forman and Godron, 1988: 83–120).

Corridors

We can define corridors as 'narrow strips of land which differ from their surroundings' (Forman and Godron, 1988: 123). They can be isolated strips, but are usually attached to a patch of somewhat similar vegetation. Two different types of corridors can be distinguished in the study area: *disturbance corridors* and *environmental resource corridors*.

The following disturbance corridors are present in the area: roads, fire lanes and a power line (Inga-Shaba). In general these phenomena are rather long, but their width is mostly smaller than the ground resolution of the SPOT multispectral imagery pixels (20 m × 20 m). These line structures consist of mixed pixels. Due to the alignment of these pixels they are visible on the original SPOT imagery. These corridors can be detected by using a series of hard filters: the so-called 'line detectors'. It allows one to distinguish small tracks which are hardly visible on the colour composite. Differentiation can be made between large corridors and minor ones by enhancing the contrast of these images. In the test site it is possible to map the important roads, paths and fire lanes in an automatic way.

Environmental resource corridors cannot be mapped in an automatic way. The stream corridors in the area consist of gallery forest and palm trees. The dense vegetation is situated along the rivers at the lower part of the slopes. The different vegetation types can be mapped by a supervised image classification. In this way it is possible to discriminate the palm trees and the gallery forest (Figure 12.3). Some stream corridors are interrupted. A field control revealed that on these places the vegetation is cut (washing places) or burned (fire raising).

Similar to patches, the stability of and the species dynamics within corridors vary widely according to its origin. Width, narrows, breaks, nodes (different types), curvilinearity, circuitry and connectivity control the important conduit and barrier functions of a corridor (Forman and Godron, 1988: 121–155).

A detailed analysis and measurement is possible on processed satellite images. Network analysis (intersection types, mesh size, α index and γ index)—as a primary measure of corridor structure—can also be performed on treated satellite data.

Matrix

The matrix can be defined as 'the extensive, relatively homogeneous landscape element that encloses scattered distinct patches or corridors of a different type' (Forman and Godron, 1988: 157). The matrix is usually the most extensive and connected landscape element type and therefore plays the dominant role in the functioning of the landscape (flows of energy, materials and species).

In the area the matrix consists of only one class, i.e. the pseudo-steppe, characterized by a sparse grass vegetation (low LAI) (Figure 12.3).

It is possible to calculate the matrix porosity—a measure of the density of patches in a landscape—on satellite images. The porosity is a clue to the degree of species isolation present and to the potential genetic variability present within populations of animals and plants in a landscape. It should also be an index of the amount of edge effect present, and relevant to animal foraging (central place foraging). Satellite images also permit one to study the boundary shapes (concave or convex): according to the 'form-and-function' principle, the interaction between two objects is proportional to their common boundary surface and type (Forman and Godron, 1988: 168–177).

Landscape change

A multi-temporal analysis of processed satellite images permits the construction of variation curves which describe variation of landscape characteristics (e.g. patch

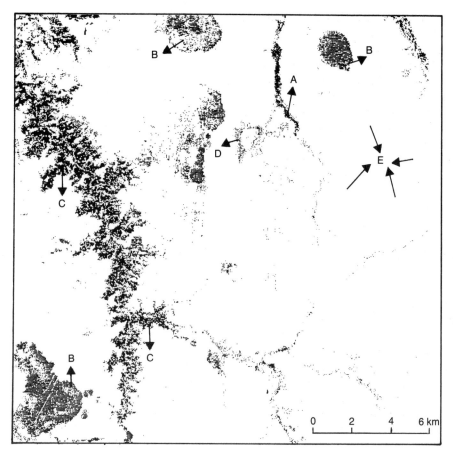

Figure 12.3. Mapping of some ecological structures in the Kwilu region, Zaire (SPOT XS 1988, band 1–2–3, half resolution, minimum distance to mean classification). A, stream corridor—gallery forest; B, environmental resource patch—open forest; C, stream corridor—palm trees; D, environmental resource patch—woodland savannah; E, matrix—pseudo-steppe.

shape, corridor width, matrix porosity, etc.) as a function of time, by three independent parameters: general tendency, amplitude of oscillation, and rhythm of oscillation. Considering two images of the 'same' landscape taken within a certain time interval, and by looking at how each landscape element is maintained or replaced by another type of element, one can construct a transition matrix and calculate the replacement rate: the percent change for each of the conversions, based on the total number of points observed. Those rates are important because changes in the structure of a landscape are related to changes in its functioning (Forman and Godron, 1988: 428–445). A multi-temporal image classification makes it possible to follow the rate of regeneration of the vegetation affected by fire raising.

As a case study the focus is here on an open forest. On the images of 1987, the forest is clearly burned and a vast part of the forest is classified as fire patch. Using the same classification method on the data of 1988, only a part of the forest is classified correctly. The part, that most affected by fire, is classified as steppe. This can be explained by the low degree of soil coverage because in this area the vegetation is not yet recovered. A follow-up for several years enables an estimation of the damage caused by the forest and steppe fires.

Belgium (Kempen)

Environmental settings

The second test site is the Kempenland in the north-eastern part of Belgium (Figure 12.4). It is a rather flat area characterized by sandy soils where dune formation has taken place during the glacial periods. Originally this region was completely covered with oaks. The oak wood was cleared in a very early stage, and it was slowly altered in a moor landscape. In the late eighteenth century a vast part of this moor was planted with pines due to a lack of fuel and mine wood. During the nineteenth century many parts were (again) cultivated due to the appearance of artificial fertilizers.

As a result of the above changes, five landscape units can be distinguished:

1. an old agricultural land characterized by open arable fields;
2. a small-scale complex of arable land and pasture, characterized by an irregular shape of the parcels—many of these parcels are bordered by hedges and trees;

Figure 12.4. Localization of the test site in Belgium.
(Reproduced with permission from Monkhouse (1974: Figure 6.1).)

3. a large-scale complex of arable land and pasture, with regular parcels bordered by hedges and trees;
4. units with small, long parcels mixed with small poplar planting; and
5. moor and woods.

Detection and mapping

Introduction

The detectability of two different features, 'land blocks' and linear elements, is investigated. From a bird's-eye view, one seems to recognize parcels in the landscape on the satellite images. A closer look at the SPOT images, however, shows that some parcels cannot be distinguished from each other, e.g. parcels with the same land use separated by barbed wire, furrows, etc.

The ability to recognize parcels is a matter of contrast between them. Therefore, the term *parcel* detection in the context of satellite image interpretation should not be used, and the new term *land blocks* is introduced: a surface surrounded by at least three linear elements. As linear elements we consider rows of trees, hedges, drainage ditches, roads, etc., or a combination of these ecological corridors.

All these patterns and structures can be detected because of clear differences in contrast, which can be enhanced by filters. The contrast can be caused by many different factors.

1. The occurrence of *hedgerows*—these linear elements are, in general, too small to be detected on the SPOT multi-spectral images as pure pixels. Differences in their reflection and their shadows, however, can generate mixed pixels.
2. Differences in *land use* and/or the *phenological stages* can cause differences in the spectral reflection.
3. *Drainage ditches* can sometimes be detected due to the effect of the groundwater table.
4. *Roads* and *paths* can cause the occurrence of some mixed pixels by their strong contrast with the vegetation.

In general, the corridors or the linear landscape elements can be detected due to their effects on spectral signature. In most cases, however, they occur as mixed pixels.

Results and discussion

Topographical maps at a scale of 1 : 10000 and updated during field surveys are compared with the edge-enhanced SPOT multispectral image in order to evaluate the possibilities and restrictions of satellite data for the detection of corridors and patches (land blocks), in relation to their size and shape.

The results can be summarized as follows. *Land blocks* with an area of more than 3 ha are unmistakably detectable on SPOT imagery; those with an area smaller than 1.2 ha cannot be detected. A transition zone occurs between 1.2 and 3 ha (Figure 12.5). As far as the length is concerned one can say that the smallest size must be 300 m. When the length is smaller than 130 m it is not possible to detect the land blocks. A transition zone occurs between 130 and 300 m (Figure 12.6). Concerning the width, the smallest size must be 120 m. Land blocks cannot be

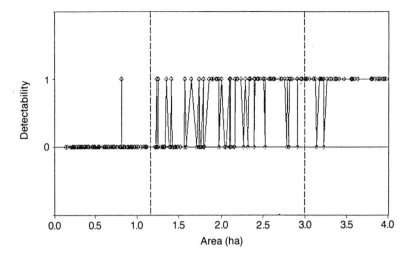

Figure 12.5. Detectability of land blocks in relation to their area. 0, Not detectable; 1, detectable.

detected when the width is smaller than 80 m. A transition zone occurs between 80 and 120 m (Figure 12.7).

In terms of SPOT XS images, one can say that a block must be built up by 28 pixels. The edges of the blocks will normally consist of some mixed pixels. This means that only 10 of the 28 pixels are pure pixels.

Figure 12.8 gives an overview of the detectability of land blocks in relation to their area and their length/width ratio. Two major groups can be distinguished: on the one hand, land blocks larger than 3 ha (group 1), and on the other hand those with an area smaller than 3 ha (group 2). All land blocks of more than 3 ha can be detected (class 1B) except those with a length/width ratio exceeding 7 (class 1A). The

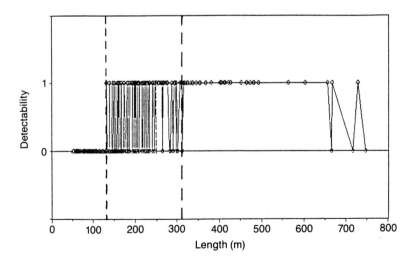

Figure 12.6. Detectability of land blocks in relation to their length. 0, Not detectable; 1, detectable.

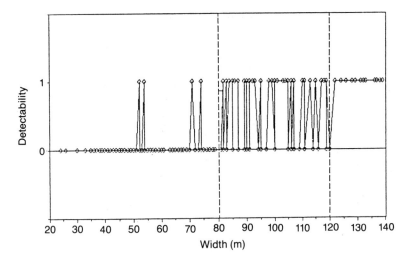

Figure 12.7. Detectability of land blocks in relation to their width. 0, Not detectable; 1, detectable.

lack of contrast due to an uniform land use plays also an important role for the undetectability of the land blocks in the class last mentioned.

Two trends can be seen in group 2 (<3 ha). (1) Land blocks with a length/ width ratio larger than 4 are generally not detectable (class 2B). (2) Land blocks with a length/width ratio smaller than 4 form a transition zone (class 2A): two-thirds are visible, and one-third cannot be detected. It must be mentioned that classes 1A and 2B do not count enough elements to be statistically relevant, they only give an idea about possible relationships and correlations. Further investigation is necessary to accept or to reject the stated hypotheses.

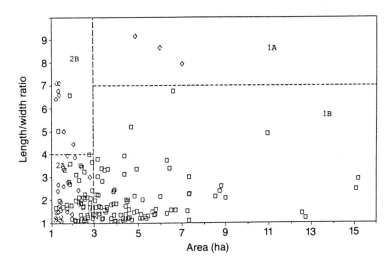

Figure 12.8. Detectability of land blocks in relation to their area and length/width ratio. (◇) Not detectable; (□) detectable.

*Table 12.2. Comparison between the results of a network analysis sustained on topo-
graphical maps and SPOT edge enhanced images.*

Connectivity index	TOPO-MAP	SPOT general	SPOT very hard filter
γ	0.98	0.98	0.69
α	0.93	0.97	0.10

All visible *linear features* are studied around land blocks of more than 3 ha.
The aim is to examine the influence of the composition on the detectability:

1. drainage ditches are clearly visible due to their influence on the groundwater
   table;
2. roads are detectable because of their high contrast with the surrounding
   vegetation; and
3. combined linear features (such as roads and ditches bordered by trees) can
   easily be detected due to the presence of the trees—when trees are replaced
   by hedges some difficulties may occur.

In summary, three linear features are important: roads, ditches and trees.
Hedges give some difficulties. On one test site, with clearly detectable land blocks, a
network analysis was performed. A comparison between the results sustained on
maps and those sustained on some edge enhanced SPOT image is made (Table
12.2). Two different 'edge' filters are used: a normal edge filter and a stronger one
('general filter'). Both filters give good results. A third filter, a very hard one, gives
very poor results.

Conclusions

The use of remote-sensing documents for the development of a global information
package for landscape ecological research can provide relevant maps. These documents are a basis—even when topographical maps of the study area already exist—
for further detailed research by ecologists. The applied methodology is suitable for
the analysis of different environments at various scales.

The methods discussed are especially useful for areas in developing countries
where drastic landscape changes can occur in a short time-span by deforestation
and uncontrolled burning of the steppes. This often leads to a degradation of the
landscape. Multi-temporal remote-sensing data can be used for studying this rapidly
changing endangered environment.

Processed and classified imagery, as discussed above, are also complementary
documents, in addition to topographical maps which are not always up to date.
They must be seen as an inventory stage for ecological monitoring on a detailed
scale.

References

Curran, P. J., 1984, *Principles of Remote Sensing*, London: Longman.
De Schlippe, P., 1986, *Ecocultures d'Afrique. Shifting Cultivations in Africa*, Nivelles:
Terres et Vie.

D'Haluin, E., 1988, 'Studie van landschapsstructuren aan de hand van SPOT-beelden (toegepast op de Zuiderkempen)', unpublished thesis, Laboratory for Regional Geography and Landscape Science, State University, Ghent.

I²S Manual, 1987, *System 600 Command Reference*, Vol. 1, Version 2.0, Milpitas: International Imaging Systems.

Forman, R. T. T. and Godron, M., 1988, *Landscape Ecology*, New York: Wiley.

Goossens, R., Ongena, T. and De Dapper, M., 1988, The use of SPOT images for the detection of soil drainage and land degradation in a wet-and-dry tropical environment (Lubumbashi, Shaba, Zaire), in *Proceedings of the 5th Symposium ISSS-WGRS, Budapest*, pp. 79–86.

Lepersonne, J., 1974, *Notice Explicative de la Carte Géologique du Zaire au 1/2 000 000*, Kinshasa: Department of Mines, Direction de la Géologie.

Mather, P. M., 1987, *Computer Processing of Remotely Sensed Images*, New York: Wiley.

Monkhouse, F. J., 1974, *A Regional Geography of Western Europe*, 4th Edn, London: Longman.

Nicolaï, H., 1963, *Le Kwilu, Étude Géographique d'Une Région Congolaise*, Brussels: Cemubas.

Richards, J. A., 1984, *Remote Sensing Digital Image Analysis. An Introduction*, Heidelberg: Springer-Verlag.

Richardson, A. J. and Wiegand, C. L., 1977, Distinguishing vegetation from soil background information, *Photogrammetrical Engineering and Remote Sensing*, **43**, 1541–1552.

Sys, C. and Devred, R., 1958, *Carte et Texte Explicative des Sols et de la Végétation du Congo Belge et du Ruanda—Urundi (1/1 000 000)*, Brussels: Ineac.

Vink, A. P. A., 1983, *Landscape Ecology and Land Use*, New York: Longman.

PART V

Applications

13

Managing environmental radioactivity monitoring data: a geographic information system approach

I. Heywood and S. Cornelius

Introduction

The landscape is an essential element to consider when determining the impacts of pollution on human health and welfare. The effects of pollutants have generally been considered in relation to individual facets of the landscape, for instance soil, water or wildlife. For pollutants for which pathways through the environment have been identified, as in the case of radioactive pollutants, a more thorough consideration of landscape ecology is essential. However, current monitoring programmes for radioactive pollutants still concentrate on the measuring of individual isotopes in isolation from other landscape elements. Geographical information systems (GIS) offer one method for combining such monitoring data with other environmental and socio-economic data to give a better picture of the role of the landscape in determining the consequences of pollution.

The accident at Chernobyl in April 1986, and the subsequent deposition of radioactivity over large areas of Northern Europe, highlighted the lack of understanding of landscape processes in determining the impact of the accident on the environment. The British Government came under severe criticism at the time for not having in place a satisfactory emergency system capable of predicting accurately such issues as the consequences to human and environmental health, and what, if any counter measures were necessary. The decisions, restrictions and policies implemented to prevent unnecessary public exposure illustrated an incomplete assessment of the importance of the landscape character and ecological processes in determining the fate of the isotopes released. These criticisms compounded those that had already previously been voiced at the Public Inquiries into Sizewell B and Dounreay (Layfield, 1987), in the discussions relating to the apparent clustering of rare cancers around nuclear generating and separation plants (Black, 1984; Craft and Openshaw, 1985), and over the long-term health effects resulting from the 1957 Windscale fire (Urquhart, 1983). A demand for more and better information to support decision-makers has followed.

In Britain, a new initiative, the National Response Plan, and its associated monitoring programme, will generate a plethora of useful radiological information, but unless steps are taken to collate, process and integrate the information collected with other environmental and socio-economic data sets it is unlikely that the full benefits will be gained and an adequate basis established for efficient radiological decision-making. Adopting GIS techniques to manage the radiological data generated by current and future monitoring programmes could enhance their value by allowing this integration with other relevant spatial data sets. This would create a new information resource which could provide information for:

1. the fine tuning of predictive dose assessment and accident simulation models;
2. use in multivariate exploratory data analysis; and
3. the rationalization and optimisation of existing routine and emergency monitoring programmes.

An overview of the current British approach to environmental radiation monitoring is presented here, followed by a discussion of the major issues which would have to be considered in formulating a GIS for the management of radiation monitoring data. Finally, examples illustrating the use of spatial data handling and automated cartographic techniques are provided from work undertaken by the authors. These examples are discussed in the context of developing a National Radiological Spatial Information System (NRSIS) demonstrator utilising GIS technology.

British monitoring strategy

Overall coordination of environmental monitoring programmes in Britain is the responsibility of the Department of the Environment (DoE) Steering Committee for Environmental Monitoring and Assessment. The current aims and objectives of the UK monitoring programme are summarized in *DoE Pollution Report No. 17* (Department of the Environment, 1983) and in the more recent Her Majesty's Inspectorate of Pollution (HMIP) report (HMIP, 1987). The principal objective, which is to assess public exposure to radioactivity, stems directly from the statutory requirements of the Radioactive Substances Act 1960. It is a current requirement of authorization by the Secretary of State that all radioactive discharges by operational authorities are kept as low as reasonably practicable and that monitoring should be implemented (if considered necessary) to ensure that this is the case.

Responsibility for the discharge of radioactive effluent as part of the UK nuclear power programme lies with the Central Electricity Generating Board, the South of Scotland Electricity Board, British Nuclear Fuels Limited (BNFL), the United Kingdom Atomic Energy Authority and Nuclear Industries Radioactive Waste Executive. The monitoring programmes of these agencies are required to provide data in two areas. Firstly on the type and quantity of the various radioactive isotopes their institutions discharge (this is usually undertaken by monitoring at source), and secondly, unless the radioactive discharge is extremely small, they are required to monitor levels within the local environment. Only selected environmental media are monitored. Some media are chosen because they are indicators of the deposition patterns and rates of accumulation of discharge related isotopes, whilst others are monitored because they form part of a pathway by which man is indirectly or directly exposed to the discharged effluent. Modelling techniques are then used to make dose-assessment calculations based upon a prior knowledge of the behaviour of isotopes within the system.

In order to optimize monitoring programmes only critical pathways and nuclides are sampled regularly. This concept of critical pathway analysis is fundamental and underlies virtually all monitoring programmes in Britain. Table 13.1 provides examples of selected critical pathways around British nuclear installations. The majority of data accruing from these programmes is made available only through publications by concerned authorities such as the DoE, HMIP and the Ministry of Agriculture, Fisheries and Food (MAFF), or in reports by the operating authorities (HMIP, 1987). The data are usually presented in an aggregated form and with only limited spatial referencing (see Table 13.2).

Organizations empowered by government to provide assurances that current methods employed by operational agencies to regulate their radioactive discharges are adequate and comply with the international and national safety limits form the second component of Britain's monitoring strategy. HMIP and MAFF are responsible for this and their main tasks include providing a way of checking for inadvertent or undocumented releases; formulating an independent database on environmental radioactivity levels; estimating public exposure; and identifying the major transfer mechanisms. Like their counterpart programmes undertaken by operational authorities, these strategies rely on the use of environmental indicators and the concepts of critical pathway analysis. The monitoring programmes instigated by HMIP and MAFF are either installation specific (e.g. the monitoring of the aquatic and terrestrial environments around all domestic nuclear facilities by MAFF) or are in the form of wider surveillance programmes (e.g. the drinking water monitoring programme and the NRPB's Environmental Radioactivity Surveillance Programme (HMIP, 1988)).

Table 13.1. Some critical pathways and critical groups identified in the UK.

Site	Pathway	Nuclide	Group
Dounreay	Contaminated fishing gear by particulate material	Cerium-144	Fishermen's hands
	Suspended matter carried into rocky clefts on coastline	Various	All those who use the beaches
Springfield	Grazing animals	Iodine-131	Those consuming local milk—principally young children
Sellafield	Harvesting locally grown seaweed (porphyra) to produce lava bread	Ruthenium-106, cerium-144	Local lava bread consumers, critical organ is the gastrointestinal tract
	Contamination of foreshore	Ruthenium-106, zirconium-95, niobium-95	All those who use the beaches
	Sea-to-land transfer	Plutonium 231	Increase in ambient dose rates in coastal locations
	Locally caught fish	Caesium-137, caesium-134	Local fish consumers
Bradwell	Oysters caught locally in the Blackwater estuary	Zinc-65, silver-110m	Local consumers of oysters
Trawsfynydd	Trout and perch	Caesium-137, caesium-134	Local consumer of fish and fishermen

Table 13.2. *Spatial referencing of terrestrial environmental monitoring carried out at BNFL Chapelcross.**

Sample/site	Frequency
Milk from six farms within 3.2 km and within 3.2–6.4 km	Each farm sampled twice monthly. Tritium analysis carried out on all samples. Analysis for other isotopes carried out on 'critical farm'. Inner bulk, outer bulk once per calendar quarter
Green vegetables from within site perimeter fence	Second and third quarters
γ dose rate—one location at nearest habitation	Monthly by TLD or film badge

* Adapted from HMIP (1988).

The final component of the existing monitoring programme in Britain is the contribution made by independent research institutions. The studies carried out by these organizations are usually area specific case studies which cover only a limited time period. Some are undertaken on contract from the DoE or on subcontract from other authorities. Although, independently, such studies may contribute significantly to the increased understanding of radioactive isotopes in the environment they can in no way be regarded as monitoring programmes. However, they very often provide information at a much finer spatial and temporal resolution than official monitoring and, therefore, can be used to complement these broader scale programmes.

The national response plan: a new coordinating initiative

In 1988, in response to the criticisms concerning the provision of information at the time of the Chernobyl accident, a new component was added to the British monitoring strategy. The announcement of a National Response Plan (NRP) and Radioactive Incident Monitoring Network (RIMNET) by the Government (HMIP, 1988) has shown recognition of the problems of data management and the necessity of access to data from all types of suppliers in the event of an incident overseas.

RIMNET is at the centre of the proposed NRP and will consist of a series of continuously operating radiation monitors capable of detecting independently any radioactive deposition in the UK. The initial proposal is for 80 monitoring stations. If anomalous radiation levels are detected by the monitoring ring then the Government requires that the NRP should be capable, very rapidly, of providing an initial assessment of the possible consequences to the UK. The responsibility for undertaking this assessment and instigating a subsequent plan of action has been placed with HMIP. In the event of another major incident HMIP will be required to provide public information and advice on any counter-measures that may be necessary.

When RIMNET becomes operational it will be the first monitoring based early warning system in the UK and in that respect is an important step forward. Careful thought must be given to its value as a predictive tool on which to base the implementation and provision of radiological advice. A considerable amount of attention has centered on this issue. The principal criticisms are from local authorities who claim that the NRP is 'all right as far as it goes' (National Society for Clean Air, 1988). They are concerned that the NRP will not provide enough information for any meaningful evaluation of radiological consequences at a regional level; for example, for crop screening, critical pathway or accurate dose-assessment calculations. Local authorities, with Government endorsement, have formulated a series of regional monitoring programmes designed to provide data at a much finer resolution to complement the RIMNET data (National Society for Clean Air, 1987). A total of 120 local authorities were in the process of establishing such programmes in 1988 and a coordinating committee has been established under the auspices of the NSCA (National Society for Clean Air, 1988).

The NRP and the local authorities realize the need to collate information into a central database facility (CDF) using commercial electronic mail (such as British Telecom Gold, Telex and facsimile) and services such as Ceefax and Oracle to disseminate information (HMIP, 1988). There is, nevertheless, considerable debate over who will have access to data and how it will be processed and used. Local authorities are particularly concerned that whilst it might be politically acceptable for them to supply information to the CDF, accessing this information might well be difficult because of the problems of confidentiality inherent in the nuclear industry.

The GIS approach

The new move toward information provision and decision support heralded by the NRP is a clear indication of the need for a data management and information system that is capable of:

1. handling the large volumes of data generated by monitoring programmes at different geographical and temporal scales;
2. providing a dynamic environment for evaluation and prediction of both the immediate and long-term consequences of the radiation levels identified in the environment;
3. allowing for the presentation of radiological information in an uncomplicated and easily understood form for easy dissemination; and
4. providing a credible methodological framework within which to increase our understanding of the behaviour of radioisotopes in the environment.

Whilst traditional aspatial management and modelling techniques provide one opportunity, adopting a GIS approach is far more flexible. There are, however, several key areas that need to be considered before such a system can become fully operational. These are the organizational and operational frameworks under which the system will exist, the data requirements and integration with other software.

An organizational framework for NRSIS

A spatial information system will only be successful if it enhances the ability of the policy-maker to make decisions in research, planning and management (Smith *et al.*, 1987). In the past one of the reasons why information systems have failed in this respect is because those developing the system have neglected the necessity for its incorporation into the organizational structure of the decision-making process (Masser and Campbell, 1988). The success of NRSIS would therefore depend on the careful integration of the system into any, existing or proposed, strategy for the coordinated monitoring programme. One way to proceed would be for NRSIS to be subsumed into the data acquisition and dissemination network being established for the NRP. A second option would be the establishment of a completely new operational framework, designed purely for coordinating the acquisition and management of all radiological information and its dissemination to relevant policy-makers (of which the agencies involved in NRP would be an example) or interested individuals (e.g. academics, the media and the general public). The responsibility for coordinating such a programme might fall within the broad remit of the NRPB. However, it is doubtful whether this would be considered politically acceptable because the Government has previously rejected a suggestion from the Royal Commission on Environmental Pollution that the NRPB should play a more central role in coordinating radiation protection strategies (see CMND, 1976, 1977). A final option might be to use existing data archive facilities such as those managed by the Economic and Social Research Council, for example the rural areas database (Lane, 1988). The principal problem with this suggestion is that, whilst it might work well with radiological data generated by academic studies, it is unlikely that either the designated authorities or operational agencies would approve. The following discussion centres on the adoption of NRSIS within the framework of the NRP because this is the strategy which the authors consider to be the most likely at the present.

A proposed operational structure for NRSIS

NRSIS should incorporate the benefits of recent advances in the development of on-line information systems and 'stand alone' workstations. The system needs to be capable of both accepting and validating data from a variety of sources, then processing these data and providing a radiological information service which will aid in the formulation of national and regional radiation protection policy. It should be capable of operating under both normal and emergency circumstances and, in the case of the latter, be able to accept, process and provide radiological information in real time. The system suggested has a modular structure which is designed to make optimum use of existing facilities and minimize the need to duplicate data storage. Several of the components in the complete system are not yet available, but the basics are already feasible and can be used to advantage, whilst also anticipating a growth in the provision and availability of spatial data over the next decade.

At the core of NRSIS should be a central database facility (CDF). This should be designed to act as the principal archive for all radiological data generated by the collaborating agencies. The system should make use of existing and future advances (such as ethernet) in electronic mailing to provide an on-line information system via which data can be deposited (or accessed). This central database need not hold information other than spatially referenced radiological data.

The second modular component of NRSIS should be provided by the regulatory monitoring agencies. This should take the form of an enhancement to their existing computer facilities to allow for the spatial data processing of subsets of data acquired from the CDF and other spatial information systems. The geographical scale of inquiry which might be required cannot be predetermined, therefore, any enhancement of the existing data-processing facilities must be sufficiently powerful and dynamic to permit investigations to focus on national, regional and local problems. The system should also be capable of supplying information to the CDF either in its raw form as monitoring data or in a value-added form after it has been subjected to spatial data processing.

The third modular component of NRSIS should be a statutory requirement, attached to the authorization to discharge toxic waste, that necessitates operational authorities to provide radiological information direct to the CDF. The extent to which these agencies might wish to adopt a GIS approach to coordinating their own monitoring programmes is for them to consider.

The radiological data-management programmes adopted by local authorities should make up the fourth module of NRSIS. These programmes should make provision for stand-alone workstations which are capable of providing regional radiological information for inclusion into the central database, and allowing access to this information as and when required. In addition, these workstations should support a geographical database containing other regional information pertinent to radiological protection. The resources required to maintain these workstations could be reduced, where local authorities are adopting GIS for coordinating regional planning purposes, by the integration of these tasks. This would also reduce the duplication and cost associated with obtaining other relevant spatial data sets.

The fifth module of NRSIS should be the provision of access to other accredited institutions to allow for the entry of data generated by research programmes and the acquisition of data from the CDF to enhance the interpretation and formulation of future programmes. A final module should be developed which would permit rapid public access to relevant radiological advice in the unforeseen event of another nuclear accident. Such a system should be designed to make use of the current state of the art electronic mailing systems.

Radiological data requirements

The mechanism for the provision of radiological data for NRSIS is already in place, due to the variety of statutory and regulatory requirements and investigative programmes which necessitate the collection of radiological information. The principal problem with these data is the lack of accurate spatial referencing. Currently, there are only two nationally agreed guidelines for standardization on spatial referencing—the Ordnance Survey (OS), National Grid, and the postcode (Department of the Environment, 1987).

The use of the postcode is only relevant where radiological surveys relate to household monitoring programmes such as the NRPB's study of radioactivity concentrations found in household dust (Fry *et al.*, 1985). The use of grid referencing satisfies the requirements for the spatial referencing of point specific monitoring programmes. However, neither of these systems provide a solution for the larger amount of monitoring data which are not point specific and are collected, for example, to assess the radionuclide concentrations in animal derived food products or in other farm produce such as crops and vegetables. The data generated by these programmes requires the adoption of some form of zonal referencing system. There are, at present, no nationally agreed standards for the geographical zoning of environmental data, but existing geographical zones such as census districts or an entirely new spatial unit could be used. What is really needed is for environmental monitoring agencies to agree on some national standard to ensure future compatibility of environmental information and avoid the current dilemma faced by those attempting to integrate social and economic information which is aggregated into over 50 different spatial units (Openshaw *et al.*, 1987).

There are currently three other sources of radiological information. Firstly, a wealth of radiological data already exist and these provide vital background information for NRSIS. The inclusion of historical radiological data in NRSIS is, however, complicated by the lack of spatial referencing. This can be alleviated to some extent where named locations are used, or where future monitoring will take place at the same location.

Secondly, there are a large number of case studies which have been carried out to investigate the behaviour of radionuclides within the environment (e.g. Howard, 1987). This type of work provides important data snapshots for various geographical areas and at differing points in time. Storing past and future case-study research would provide a further invaluable source of information. The potential and techniques for storing this type of 'data snapshot' information have already been investigated; for example, the BBC Doomsday Project (Rhind and Openshaw, 1987) and similar techniques could be adopted as an extension to NRSIS.

Finally, advances in airborne radiometric surveying techniques mean that radiological data can now be collected in real time. Moreover, because these data are captured in digital form, GIS techniques provide an excellent tool for the immediate processsing of this information. This has numerous implications for both emergency monitoring and the formulation of subsequent emergency plans.

A problem faced by developing the radiological component of NRSIS is the issue of reliability. This is particularly relevant because NRSIS would provide wide access to any radiological data contained within the system, and these data may well be used in management decisions. Therefore, it is essential that within the data-acquisition section there must be a verification system. The system of accreditation being established by the NRPB for inclusion of laboratories in the RIMNET monitoring programme provides an ideal standard from which to develop a framework for NRSIS. Some form of quality control must also be established for historical data. The attachment of an error rating to the data so that those accessing the data are aware of its reliability may be one solution.

Non radiological data for NRSIS

For effective decision-making it is essential that other topographic, thematic and social data sets are available to NRSIS. The range of relevant environmental spatial data sources is growing rapidly (National Environmental Research Council, 1983) and any future radiological data-management system needs to make provision for its incorporation. What is eventually included will depend not only on what is available, but also on the real requirements of those planning to use the system.

Access to wide sources of digital data will open up new opportunities for the use of techniques such as map overlay in conjunction with dynamic modelling. For example, after identifying localized source areas for the uptake of radionuclides by plants, and subsequently animals, agricultural census data could be used to assess the real importance of the levels identified in terms of local food production, with social and demographic data sets helping to identify the local population most at risk.

A number of other on-line information systems are currently being developed to provide a variety of up to date environmental information. One example is the CORINE programme of coordinated information on the European environment (Wiggins *et al.*, 1987). These systems, like NRSIS, are being designed to provide access to information in real time and, therefore, any future development of NRSIS should consider these systems as potential data sources.

Integration with existing data-processing software

In order for NRSIS to make the best use of existing resources there is a need for the system either to incorporate existing data-processing software, or simply to act as a data library providing geographically specific information for use in existing abstract models. The former option is an extended version of the latter and would work well for those data-processing techniques which use only simple mathematical and statistical techniques. The conversion of regional radiation activity levels into generalized dose equivalents is a prime example. However, where substantive mathematical models already exist (see, for example, Linsley *et al.*, 1982; Nair, 1984) it would be a waste of resources to redevelop these within a new GIS environment. Moreover, GIS software is not yet developed enough to provide a sophisticated framework for the development of complex mathematical models (Openshaw, 1988). In this instance, NRSIS could provide the data libraries through which existing complex models can gain access to large sources of data. In the past, the lack of available data has been a recognized problem faced by those involved with the development and validation of these abstract models (Jackson and Smith, 1987). If NRSIS is to prove a flexible tool it must be capable of recapturing the value-added data generated by these models.

The advantages of adopting NRSIS

Some of the advantages of using NRSIS can already be identified. For example, NRSIS could provide:

1. rapid access to radiological and other relevant information for use in routine and emergency radiological protection decision-making;
2. routine updating of both radiation monitoring data and associated geo-coded data sets;
3. addition of new spatial data sets at such times as they are considered relevant or become available (e.g. data from airborne radiation monitoring surveys);
4. comparison of different data sets to investigate a wide range of spatial and temporal relationships;
5. verification of external modelling programmes;
6. inclusion of data stored in a variety of other forms and locations;
7. integration with other national and international environmental information systems;
8. identification of (a) critical pathways and (b) areas which because of the unique character of their physical environment may act as both sinks and sources for radioisotopes;

9. dissemination of radiological information in a wide variety of forms; and
10. optimization of existing national and regional monitoring programmes.

At this stage it is all too easy to become entranced by the obvious potential benefits of melding together a wide range of interrelated data. There are, however, a whole host of issues, principally associated with availability, accessibility and compatibility that pose problems for the development of large-scale GISs. These have been discussed at great length by a number of authors (e.g. Smith *et al.*, 1987; Rhind, 1987; Estes *et al.*, 1987) and broadly fall into two categories: technical and organizational. Technical problems mainly focus on the issues surrounding compatibility and the need for standards for locational referencing, choice of data structures and methods of data acquisition. These technical problems are considered to pose only short-term barriers (Department of the Environment, 1987). The organizational problems comprising the possessive attitudes towards information and the genuine problems of confidentiality and copyright pose a more serious problem and require a fundamental change in thinking over data ownership. However, the recent agreement reached with the utility companies over the provision of large-scale maps (Chorley, 1988) is indicative of a move towards a solution. These problems need to be rectified at a National and Regional scale if GIS applications are to go ahead in a number of areas and are, therefore, not unique to the development of NRSIS.

The NRSIS demonstrator

Questions raised by north-eastern local authorities over the handling, management and dissemination of radiological information (Anderson *et al.*, 1987) were the initial inspiration for developing the NRSIS demonstrator. Local authorities were particularly concerned over how they should go about designing both routine and emergency regional monitoring programmes to make the best use of their scarce resources whilst still providing valuable radiological information. The NRSIS demonstrator was developed with the aim of simulating the role that a GIS workstation would perform in aiding the collation, processing and presentation of radiological information under both accident and routine monitoring conditions. The demonstrator was developed using MicroVax and PC implementations of the ARC/INFO GIS software. Cumbria was selected as the area to be covered, primarily because of the wealth of radiological data available for this region. However, the system was designed so that it could be implemented for any region. Five objectives underpinned the programme of development.

1. To establish the format for a spatial database for the storage, retrieval and analysis of monitoring data.
2. To collect and prepare data on environmental factors (such as soil type and land use) which may have an effect on deposition patterns, and data on factors which may be affected by deposition patterns (e.g. agricultural and census data).
3. To investigate the spatial interrelationships between the radiological and environmental data sets.
4. To evaluate whether the relationships identified in stage three could be used to develop a sample site optimization routine and a regional sampling classification scheme.

5. To examine methods by which the GIS could be used to identify 'hot spots' and establish estimated dose equivalents and generalised derived safety limits for various environmental media.

Data layers

The NRSIS demonstrator makes use of a selection of physical, human and radiation monitoring data (see Figure 13.1). Two types of geocoded data have been used: zonal thematic map data and grid referenced point data from either paper-based sources or existing digital databases. Information included so far is for soils and administrative boundaries covering the Cumbrian area. The Ordnance Survey national grid referencing standard was adhered to so that other data could be added easily. Grid referenced data include rain-gauge sites and radiation-monitoring sites.

Thematic data incorporated so far have included selected items from the 1981 Population Census and the 1980 Agricultural Census. Items were selected from these databases on the basis of their importance to radiation planning. Population statistics cover such variables as total population, numbers of males, females, people under 16 years old and pensioners. General agricultural data such as areas under grass, woodland and total area have also been included, together with more specific information on sheep populations (number of breeding ewes, number of lambs under 1 year old, number of rams, etc.). These data were obtained in a digital form from the Economic and Social Research Council's Rural Data Archive. The use of the Joint Academic Network (JANET) to acquire these data are illustrative of how easily these data could be made available to local authority systems. In fact many

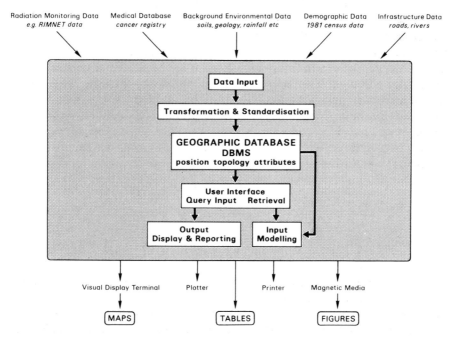

Figure 13.1. An integrated radiation monitoring and analysis system.

local authorities already make use of census data in this way via the National Online Manpower Services Information System (Townsend *et al.*, 1986). It should be noted that, although available in digital form, the data obtained in this manner required a considerable amount of preprocessing to facilitate input to the GIS package.

The information included in the demonstrator represents only a subset of the data layers which it is anticipated a completed system would include.

Data retrieval

One of the main aims of the demonstrator was to produce a 'turnkey' GIS environment via which information, pertinent to radiological protection, could be easily and quickly obtained by a relatively inexperienced user. Several different options for retrieving data have been developed, allowing for the visual display of data. In all cases the emphasis has been on keeping the methods simple, and the user is guided through a menu-driven environment supported by an informative, text-based, help system. Through the main menu the user has access to both databases and map libraries comprising the various radiation and environmental data sets, and options are available permitting the data to be retrieved in tabular or map form.

Tabular output is specified by search criteria supplied interactively by the user, in very much the same way as more traditional database management systems. The selected data file is displayed with a key and a browse facility provided so that the user can scan the information contained in the file.

The other data retrieval options use a map, chosen from the map library as the 'key' to the databases. The first of these allows the user to select a particular area of interest from a map displayed on the screen using either a mouse or the cursor keys. The query is then specified by the user and the relevant data retrieved and displayed. At any stage the user may 'zoom-in' for a clearer picture of the area of interest.

The second data-retrieval method is particularly relevant to radiological information, but could be extended to other types of data and combinations of data types. The system allows the user to select not only the particular nuclide of interest, but also the threshold level above which sample points are to be displayed. This enables individual or aggregated data from monitoring surveys to be evaluated quickly and potential 'hot spots' identified with ease. This data-retrieval method is currently being refined to allow the user to specify not only thresholds for radiological data, but also for other environmental criteria. For example, the point mapping of radiation monitoring sites for a specific nuclide and threshold level where mean annual rainfall is above a given level and the soil type of a certain character. This method of multivariate 'pinpoint' mapping helps to present a clear visual picture of regions where problems might occur, and aids the identification of the spatial interrelationships between datasets. So far this method has been used to identify the soil and agricultural characteristics associated with a particular monitoring site to aid the identification of important sites for resampling in an emergency situation.

Analysis and modelling

Despite the fact that GIS systems do not yet provide an ideal environment for the development of complex mathematical models several simple operations have been

performed using the NRSIS demonstrator which are pertinent to radiological protection. The first of these is the quick and easy way in which the NRSIS demonstrator can use basic monitoring data to develop a series of value-added maps. One
example of this has been discussed earlier in relation to data retrieval, whereby the
user could select a threshold level above or below which information is to be displayed. NRSIS can also generate new information using base monitoring data to
calculate regional dose rates and more specific information of predicted concentrations in food products. The latter can then be compared with generalized derived
safety limits to assist the evaluation of the threat to human health. One method for
this involves the interpolation of point data using geoprocessing techniques such as
kriging (Burrough, 1986). The resulting surface can be used to produce zonal maps
for administrative purposes. A series of thematic maps developed in this way is
shown in Figure 13.2.

A second analytical use to which the GIS has been put is in impact modelling
for restriction zone planning purposes. In the light of the confusion that arose after
the Chernobyl incident in 1986, it was decided to use the demonstrator to identify
where preliminary restriction zones for the movement of sheep following a similar
deposition event should be located. This was tackled by using the GIS to reselect
areas of high annual rainfall and parishes with large sheep populations from the
appropriate maps. The two maps obtained were then overlain so that those areas
containing both large sheep populations and high rainfall could be easily identified.
Although based on historical data, a good agreement was obtained with the
restriction zones implemented following the Chernobyl incident in the south-west of
Cumbria (see Figure 13.3). The exception is the area identified to the south-east of
Cumbria. This area borders the west Pennines and was not initially considered
for restriction, but was discovered by chance almost a year later to have experienced
considerable deposition. A GIS strategy would, as illustrated, have identified this
area and would prove valuable for this kind of analysis in an emergency situation.
GIS techniques would take considerably less than the 1 month required after the
Chernobyl incident, to identify these zones and a result would be available in a
matter of hours if the characteristics of the relevant critical pathways were known.

Once the data layers are in the system, it is possible to perform any other
number of useful operations, for example, calculating the population living in areas
of differing radiation levels, or the number of cattle available for slaughter or supply
to the food market. Areas of soil types known for their ability to act as sinks or
sources of radionuclides can be identified and crop yields estimated. In addition,
many new questions can be asked, for example, those pertaining to the radiation
versus health debate (Heywood *et al.*, 1989; Cross, 1989).

Conclusion

Understanding and identifying the supply, movement and distribution of artificial
radioisotopes within ecosystems is an essential requirement in formulating effective
radiological protection strategy. In Britain, routine monitoring programmes, whilst
being aimed primarily at meeting the statutory requirements of the 1960 Radioactive Substances Act, generate annually a considerable amount of valuable data on
the behaviour of radionuclides. However, the lack of a corporate approach to the
collection, management and processing of this information over the last 40 years has
resulted in the majority of these data being underused. Events following the Cher-

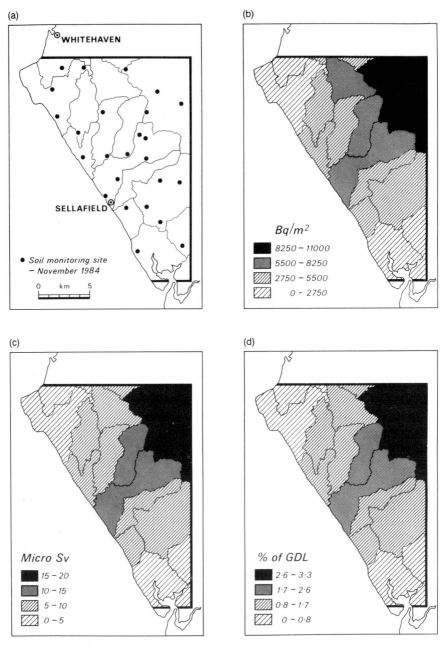

Figure 13.2. Thematic radiation maps for selected parishes in Cumbria generated from point monitoring data. (a) Soil monitoring site. (b) Levels of caesium 137 in soils. (c) External dose rates. (d) Activity levels as a percentage of the generalized derived safety limit for soil.
(Reproduced with permission from Heywood (1987).)

nobyl incident in May 1986 confirmed this, when those involved with establishing and evaluating the consequences of the accident to Britain claimed, in defence of their inabilities to provide comprehensive advice on radiological protection, that there was insufficient background information available on the behaviour of nuclides within the environment upon which to base decisions.

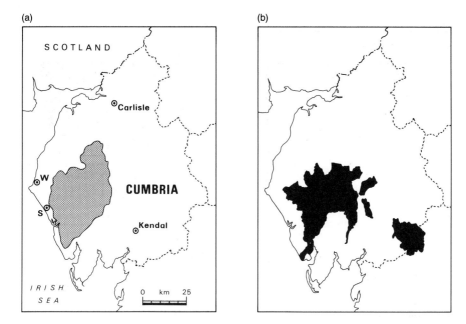

Figure 13.3. Comparison of sheep restriction zones. (a) Restricted zones as designated by MAFF, 7 August 1986. (b) Restriction zones identified using GIS techniques.

To date, this lack of information has been viewed as a data-deficiency problem rather than one of availability and coordination. The political response has been to increase funding for radiological research, to step up existing monitoring programmes, and to establish a national response programme to cater for the consequences of an incident abroad.

It has been suggested here that the concepts which underpin the National Response Plan and the implementation of RIMNET could be extended using GIS technology to make provision for a radiological protection decision-support system. The ideas behind NRSIS involve the integration of radiological information from routine, emergency and scientific monitoring programmes with other relevant spatial data sets (including land use (from Landsat), geology, soils, and agricultural and population census data) to provide for research, regulatory and emergency planning requirements. Work on the NRSIS demonstrator has shown that, once a GIS is built up, it becomes possible to develop a variety of application-specific views of the database for management purposes and for descriptive and process studies, for example, the identification of potential restriction zones or 'hot spots'. The demonstrator illustrates the potential of GIS for producing a user-friendly environment for the fast and accurate retrieval of data as well as making provision for the analysis of spatial relationships of data from different sources.

The demonstrator described in this paper is a prototype system and the NRSIS proposal is a concept with two major problems limiting further development. First there are technical problems but, as Chorley (1988) indicates these should be seen in the light of rapid advances in computer technology and artificial intelligence (Fisher *et al.*, 1988) which will increase the amount of digital data that can be handled, and extend the type and nature of complex spatial questions that can be asked. Second, there are organizational problems which underpin the development of all large-scale GIS applications, these include data availability, standardization and confidentiality, as well as those which are unique to radiation planning, particularly the need to agree on a coordinated approach to the collection and management of data. Despite these limitations the adoption of a GIS approach, could for the first time, provide a system capable of meeting the complex information requirements of those involved in radiological decision-making.

References

Anderson, N., Evans, P. R., Heywood, D. I., Holland, G. and Leat, P. 1987, 'Proposal for a Monitoring Programme for Environmental Radioactivity submitted to a consortium of Local Authorities in the Northern Region'.

Black, D., 1984, *Investigation of the Possible Incidence of Cancer, in West Cumbria*, London: HMSO.

Cambray, R. S. and Eakins, J. D., 1980, Studies of Environmental Radioactivity in Cumbria, Part 1, Concentrations of Plutonium and Caesium-137 in Environmental Samples from West Cumbria and a Possible Meantime Effect, AERE Harwell Report R9807, (HMSO, London).

Chorley, L., 1988, Geographic information systems. Preface to working with geographic information systems, *Economic and Social Research Council Newsletter 63*.

CMND, 1976, *Royal Commission on Environmental Pollution, Sixth Report, Nuclear Power and the Environment (CMND 6618)*, London: HMSO.

CMND, 1977, *Nuclear Power and the Environment, the Government Response to the Sixth Report of the Royal Commission on Environmental Pollution*, London: HMSO.

Craft, A. W. and Openshaw, S., 1985, Childhood cancer in West Cumbria, *The Lancet*, **8425**, 403.

Cross, A., 1989, 'Childhood leukaemia in northern England', unpublished Ph.D. thesis, University of Newcastle upon Tyne.

Department of the Environment, 1983, State of the art review of radioactivity monitoring programmes in the UK. A report by the radioactive monitoring group to the steering committee for environmental monitoring and assessment, *Pollution Report No. 17*, London: HMSO.

Department of the Environment, 1987, *Handling Geographic Information* (report of the Chorley Committee of Enquiry chaired by Lord Chorley), London: HMSO.

Estes, J. E., McGwire, K. C., Fletcher, A. G., and Foresman, T. W., 1987, Coordinating hazardous waste management activities using geographical information systems, *International Journal of Geographical Information Systems*, **1**(4), 359–377.

Fisher, P. F., Mackaness, W. A., Peacegood, G. and Wilkinson, G. G., 1988, Artificial intelligence and expert systems in geodata processing, *Progress in Physical Geography*, **12**(3), 371–388.

Fry, F. A., Green, N., Dodd, N. J. and Hammond, D. J., 1985, Radionuclides in House Dust, *NRPB Report R181*, London: HMSO.

Heywood, D. I., 1987, 'Environmental radiation monitoring and the siting of nuclear facilities', unpublished Ph.D. thesis, University of Newcastle upon Tyne.

Heywood, D. I., Cornelius, S. C., Openshaw, S. and Cross, A., 1989, Using a spatial database for environmental radiation monitoring and analysis, in *Proc. GIS—A Challenge for the 1990s*, Ottawa, Canada.

Howard, B. J., 1987, Cs-137 Uptake by sheep grazing tidally inundated and inland pastures near the Sellafield reprocessing plant. Pollutant transport and fate in ecosystems, in Coughtrey, P. J., Martin, M. H. and Unsworth, M. H. (Eds), *Special Publication Number 6 of the British Ecological Society*, London: Blackwell Scientific Publications.

HMIP, 1987, *Monitoring Radioactivity in the UK Environment*, London: HMSO.

HMIP, 1988, *Nuclear Accidents Overseas: The National Response Plan and Radioactive Incident Monitoring Network (RIMNET) Statement of Proposals*, London: HMSO.

Jackson, D. and Smith, A. D., 1987, Generalized models for the transfer and distribution of stable elements and their radionuclides, in Coughtry, P. J., Martin, M. H. and Unsworth, M. H. (Eds) *Agricultural Systems in Pollutant Transfer and Fate in Ecosystems Special Publication No. 6 of the British Ecological Society*, pp. 385–402, London: Blackwell.

Lane, M., 1988, The rural areas database, *Mapping Awareness*, **2**, 2.

Layfield, F., 1987, *The Sizewell Inquiry*, London: HMSO.

Linsley, G. S., Simmonds, J. R. and Haywood, S. M., 1982, FOOD-MARC: the foodchain transfer module in the methodology for assessing the radiological consequences of accidental release, *National Radiological Protection Board NRPB-M76*, London: HMSO.

Masser, I. and Campbell, H., 1988, Working with geographical information systems: the organisational context, *Economic and Social Research Council Newsletter 63*, October.

Nair, S., 1984, Models for the evaluation of ingestion doses from the consumption of terrestrial foods following an atmospheric radioactive release, *CEGB Report, RP/B/52004/N84*, Berkeley: Berkeley Nuclear Laboratories.

National Society for Clean Air, 1987, 'The Role of Local Authorities in Relation to Radiation Monitoring Arising from an Accident Outside Britain, The Department of the Environment's 'National Response Plan', a supplementary briefing prepared by the NSC, November 1987.

National Society for Clean Air, 1988, First report of the Civil Nuclear Power Working Group to the Technical Committee of the National Society for Clean Air, *Clean Air*, **18**, 2.

Natural Environmental Research Council, 1988, *Geographic Information in the Environmental Sciences*, report of the Working Group on Geographic Information.

Openshaw, S., Goddard, J. and Coombes, M., 1987, Integrating geographic data for policy purposes: some recent UK experiences, *Northern Regional Research Laboratory Report No. 2*, NRRL.

Openshaw, S., 1988, Developments in geographical information systems. Working with geographical information systems, *Economic and Social Research Council Newsletter 63*, October.

Rhind, D., 1987, Recent developments in geographic information systems in the UK, *International Journal of Geographic Information Systems*, **1**(3), 229–243.

Rhind, D. and Openshaw, S., 1987, The BBC Doomsday System: a Nation-wide GIS for $4448, *Proceedings of Auto Carto 8*, Washington DC.

Smith, T. R., Menon, S., Star, J. L. and Estes, J. E., 1987, Requirements and principles for the implementation and construction of large-scale geographic information systems, *International Journal of Geographic Information Systems*, **1**(1), 13–31.

Townsend, A., Blakemore, M., Nelson, R. and Dodds, P., 1986, The National On-line Manpower Information System (NOMIS), *Employment Gazette*, **94**(2), 60–64.

Urquhart, J., 1983, Polonium: Windscale's most lethal legacy, *New Scientist*, **97**(1351), 873–875.

Wiggins, J. C., Hartley, R. P., Higgins, M. J. and Whittaker, R. J., 1987, Computing aspects of a large geographic information system for the European Community, *International Journal of Geographic Information Systems*, **1**(1), 77–87.

14

Using hydrological models and geographic information systems to assist with the management of surface water in agricultural landscapes

R. A. MacMillan, P. A. Furley and R. G. Healey

Introduction

Geographical information systems (GIS), remote sensing, and the various analytical tools associated with them, are increasingly able to provide sources of data and means of manipulating data appropriate for large scale management of resources. A major management concern in western Canada is the conflict between the desire of farmers to drain or consolidate small bodies of non-permanent surface water and the possible on- and off-site effects of such land drainage.

This chapter describes progress in the use of digital elevation data, hydrological simulation modelling and GIS to assess the time-varying distribution of non-permanent surface water at a farm scale and to estimate the possible patterns of redistribution that might arise from drainage or consolidation. A number of alternative techniques and modelling approaches have been investigated. Emphasis has been placed on simulation models that focus on geomorphic control of patterns of surface water flow and accumulation. The geometry of surface water flow is controlled by surface topography whilst the timing of flow and the magnitude of water accumulation in depressions are related to soil, climate and land cover characteristics in addition to geomorphic form. The models adopted define the geometry of flow and attempt to simulate the magnitude and timing of flow. GIS capabilities have been found essential for acquiring, collating, storing and displaying the data needed for these models and for displaying and outputting the results.

Definition of the problem

The flow of water across the landscape and through the soil establishes spatial patterns and structures fundamental to studies of landscape ecology. Water flow is, in fact, the main integrating factor in many interdisciplinary studies of the environment. Human activities play an important role in establishing or altering existing patterns of water distribution.

A landscape pattern of current interest in western Canada is that created by the accumulation of water as non-permanent wetlands, in small, shallow, discrete

Figure 14.1. Oblique aerial photographs illustrating (a) pitted 'prairie pothole' terrain and (b) non-pitted terrain conditions desired by farmers for uniform cultivation.

depressions (Figure 14.1). Interest in these wetlands arises from concern over the growing agricultural practice of removing or consolidating ephemeral ponds to achieve more efficient farming patterns. There is, at present, no satisfactory methodology for investigating the possible effects of disruption to existing landscape patterns arising from such farm scale drainage activities.

The study described here involved investigating the potential of using some of the new tools associated with spatial information systems to help predict the possible effects of farm scale water management. The discussion presented here revolves around a detailed case study of a farm scale study area. The focus of this chapter is the background to the problem, problem definition, the overall conceptual design, the modelling approaches that have been used and the initial results that have been achieved.

Background and problem definition

It is unusual to think of excess surface water in an environment as arid as that of Alberta where annual evapotranspiration potential exceeds precipitation by 200 mm or more. Rapid run-off does occur, however, in two instances; during spring snow-melt, when the subsoil is frozen and following heavy precipitation onto soils which have low infiltration capacities or are saturated. Such run-off accumulates in the many small depressions characteristics of the 'prairie pothole' type of terrain common in the glaciated landscapes of western Canada (Stewart and Kantrud, 1969). Once concentrated in these depressions, run-off disappears rather more slowly through evaporation and infiltration. It has been recently estimated (Alberta Water Resources Commission, 1987) that of the approximately 12 million acres of wetlands in the agricultural area of Alberta, up to 2 million acres may be non-permanent wetlands of the prairie pothole type.

Political, economic and environmental interests in managing non-permanent surface water

From the farmer's perspective, undrained 'prairie potholes' are undesirable obstacles that reduce efficiency, decrease yields and increase production costs (see Figure 14.1). Recent studies (Alberta Water Resources Commission, 1987; Anderson, 1987; Leskiw, 1987) have confirmed that artifical drainage of non-permanent water bodies increases productivity and lowers production costs by:

1. increasing the total area under cultivation;
2. improving crop quality due to reductions in weed growth, lodging and frost risk;
3. increasing flexibility and opportunity in timing and management of field operations; and
4. improving field patterns and working efficiency by reducing overlap area for operations such as tillage, seeding, fertilizing and spraying.

Given the tight economics of farming, farmers are increasingly interested in removing 'excess' surface water. Off-farm drainage and on-farm consolidation of ponded surface water are amongst the management options considered. If costs for off-farm drainage and wildlife habitat mitigation are not assessed against individual farmers, net financial benefits of drainage can be up to $45 per acre per year (Alberta Water Resources Commission, 1987).

Despite the advantages offered to individual farmers, there are significant environmental costs associated with agricultural land drainage. These costs arise from problems with down-slope management of water released by off-farm drainage schemes and with mitigation of waterfowl habitat losses. The recently completed Alberta Drainage Inventory Study (Alberta Water Resources Commission, 1987) demonstrated that the net present values of drainage were negative, from a societal perspective, for every basin studied, and concluded that off-farm drainage of non-permanent wetlands was economically feasible only if wildlife mitigation costs were not considered.

Not surprisingly, local and regional governments have acquired an interest in exercising some planning control over agricultural schemes for surface water management. Local authorities are mainly interested in preventing conflicts between neighbouring farmers over adverse downstream impacts. Provincial agencies take a broader view that encompasses not only downstream impacts but also on-farm impacts such as loss of waterfowl habitat.

Local and regional authorities have two main concerns about surface water:

1. How much water is located on any given parcel at any given time?
2. What might be possible 'impact' on any proposed drainage or redistribution of surface water?

These questions may be restated as follows.

1. Where, and to what maximum extent, can water accumulate on any given farm (i.e. where and how large are the locations of 'depressional storage')?
2. Where might water flow and reaccumulate if it were removed from its initial natural storage location (i.e. possible down-slope impacts of artificial drainage)?
3. At any given time of interest, what is the approximate extent and volume of surface water (i.e. the extent to which 'depressional storage' is filled)?

The political, economic and environmental concerns with off-farm artifical drainage have led to increased support for the alternative of on-farm water management. The current view is that on-farm redistribution and consolidation might fulfil the farmer's need for increased efficiency and productivity while avoiding most of the adverse impacts of wildlife habitat loss and off-farm flooding complications.

With increased support for on-farm water management has come the realization that both the data and the techniques needed for effective management are lacking. Technical studies (Alberta Water Resources Commission, 1987; Jensen and Wright, 1987; W-E-R Engineering Limited, 1987) have recommended:

1. the development of improved hydrotechnical modelling to estimate runoff conditions, especially at the farm scale;
2. demonstration of detailed hydrological simulation modelling for an individual field- or farm-sized project using a detailed field scale hydrological model;
3. field programmes to estimate depressional storage volume for agricultural landscapes leading eventually to the development of regionalized predictive equations for wetland surface area/depth/volume relationships;
4. establishment of benchmark agricultural run-off research sites to measure rainfall, snowfall, temperature, sunshine hours and run-off information; and
5. further studies to determine the effects of spring snowmelt/rain run-off events on agriculture in Alberta.

Scientific and technical interest in estimating surface water flow and accumulation

New tools associated with GISs are offering possibilities for the collection, processing and use of information about surface water. It is now possible to obtain information routinely at the spatial and temporal resolutions at which it is needed. With these recent innovations, it is now feasible to deal with dynamic processes. In general, two broad groups of different, but related, techniques are of interest. They are remote sensing and distributed hydrological modelling.

Use of remote sensing to estimate surface water extent

Repeated monitoring through satellite or aerial remote sensing may be used to determine the changing location and extent of surface water bodies above a given minimum size which depends upon the resolution of the sensing platform. Some success has been achieved in using satellite remote sensing to monitor the aerial extent of depressional water bodies larger than 2 acres on the Canadian prairies (Koeln *et al.*, 1986, 1988). Other studies have reported considerable confusion and classification error, especially for the smaller bodies of surface water mostly targeted for farm scale artifical drainage (Alberta Water Resources Commission, 1987). Even if repeated remote sensing could supply the information on location and extent of surface water at the high spatial resolution required, there still would remain a need to estimate the volume of water bodies and to establish the patterns of flow and interconnection between depressions. Regional predictive equations could be developed to relate pond areas to pond volumes, but interconnectivity is controlled by topography and requires interpretation of topographic data to establish.

Use of simulation modelling to estimate surface water extent

Continuous simulation modelling offers the other option for acquiring the information needed for farm scale water management. Two distinct approaches to distributed hydrological modelling exist; topographically based modelling of run-off geometry and physically based, distributed modelling of run-off processes.

Topographically based run-off modelling is exemplified by a distinct group of geomorphic modellers (Band, 1985, 1986a, b; Jenson and Dominique, 1988; Morris and Heerdegen, 1988). This group concentrates on establishing the morphology and spatial patterns associated with run-off. Their interest is in defining the structures that arise from flows and in establishing the structural and geometric controls influencing flow. They are generally less interested in the timing of water flow or in accurately simulating the physical processes which control water flow. The results of their work are usually expressed as maps of static features such as catchment areas, flow paths and drainage networks.

Physically based, distributed modelling of run-off processes is advocated by a distinctly different group of hydrological modellers (Beven, 1977; Beven and Kirkby, 1979; Beven *et al.*, 1984; Rogers *et al.*, 1985; Abbott *et al.*, 1986a, b; Bathurst, 1986a, b; Rohdenburg *et al.*, 1986). They focus on time and process. They are interested in realistic simulation of the processes governing water flow and in establishing the magnitude and timing of run-off events. They generally provide less detail on the specific pathways or locations of flow or the spatial patterns of output. Their results are usually expressed as stream discharge rates displayed as unit hydrographs, generally at specified points such as gauging stations or basin outlets.

The geomorphic group of modellers is closely associated with the use of GIS and with the linking of hydrological models to GIS. They use spatial information systems to input, collate, organize and display the spatially varying data and to portray the output results. Increasingly, their hydrological modelling tools are being integrated with commercial GIS (Heil, 1980; Silfer *et al.*, 1987; Jenson and Dominique, 1988; Weibel *et al.*, 1988; Weibel and DeLotto, 1988) or custom-built GIS (Johnson, 1989; van Deursen and MacMillan, 1991). The hydrological group of modellers is less strongly associated with the use of GIS. Most of their models are presently 'stand-alone batch programs' with limited capability for interactive input, output or display of spatially referenced data.

Research hypothesis and conceptual design

The project is structured in terms of a 'testable' hypothesis which can be subjected to some form of 'critical evaluation'. The approach taken is to attempt to model the flow and accumulation of surface water at the farm scale and then to test, using independent field data, the capability of the models to predict the observed distributions. The primary hypothesis is that highly distributed hydrological modelling can be used to predict the likely patterns of distribution and accumulation of surface water in a farm scale landscape at any given time. Specifically, simulation of topographically controlled run-off using highly distributed hydrological models can describe:

1. the natural capacity of a farm scale landscape for storage of surface water in closed depressions;
2. the likely patterns of overflow and interconnection between closed depressions in a farm scale landscape; and
3. the location and extent of surface water accumulation at any given time.

The design adopted involves the following basic components (Figure 14.2):

1. characterize the geometry of flow and ponding at a single selected study site;
2. estimate the timing and magnitude of surface water flow and accumulation;
3. establish the actual extent of surface water ponding at the selected site at two or more specific times; and

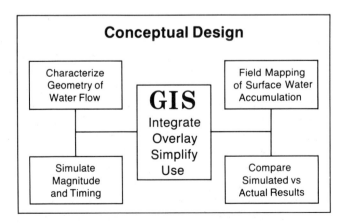

Figure 14.2. Components of the conceptual design adopted for the study.

4. compare the results of the simulated estimates of water distribution with the measured data on actual ponding.

Detailed characterization of the study site is accomplished by automatic processing of DEM data using geomorphically oriented run-off routing simulation models (van Deursen and MacMillan, 1991). These models permit the definition of flow paths and catchment areas and the complete description of the location, extent, volume and interconnectivity of all depressions. Estimating the timing and magnitude of surface water flow and accumulation is undertaken using physically based distributed hydrological models. These models permit simulation of surface water flow and accumulation which in turn allows estimation of the actual volume of ponded water at specific times. Detailed ground surveys have been used to map out the actual extent of ponded surface water at specific times. GIS software is available to help compare the results of the simulated estimates of water distribution with the measured data on actual ponding.

Methodology and preliminary results

Progress has been achieved in each of the aspects of the conceptual design outlined above. A site has been selected and the digital elevation data required to define the geometry of runoff has been acquired and processed. A simple, but highly distributed, model to simulate surface run-off and ponding in depressions has been developed. This proved necessary after a review of the hydrological and GIS literature failed to reveal a suitable existing model although elements of existing models were used to develop the final approach. Field work has been completed at the site to outline the actual distribution of water at several time intervals and to measure soil hydraulic properties required for input into the hydrological simulation models. No comparisons of simulated and observed distributions have yet been completed but some initial suggestions for comparison techniques have been proposed. A simple, raster based GIS (PC-Geostat, University of Utrecht, 1987) has been adopted to facilitate the management of spatial data.

The field site data set

The Lunty field study site is located on an actively farmed parcel of land near the town of Forestburg abut 200 km south-east of Edmonton, Alberta (Figure 14.3). The site occupies an area measuring 700 m × 800 m which is about the dimensions of a quarter section, the standard farm scale management unit in western Canada.

The site is typical of cultivated landscapes within the black-soil zone of east central Alberta that are affected by excess surface run-off. It is part of a low relief, undulating to hummocky till plain. Maximum local relief is 8 m and elevations range from 721 to 729 m. Slope gradients average 2–5% for the upper, convex portions of the landscape. Few slopes are steeper than 6–9% (Figure 14.4). Slope lengths average 75–150 m and range from 50 to 400 m. The many depressions are level to slightly convex with slope gradients of 0–2%. Depressions range from less than 0.1 to over 4 ha in size. There is no evidence of an integrated drainage pattern and surface water flows into local, undrained depressions where it accumulates and

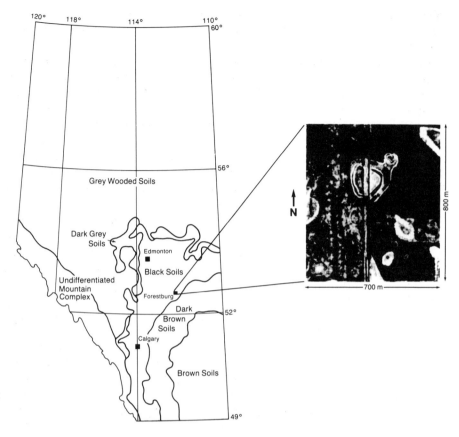

Figure 14.3. Location and extent of the Lunty site.

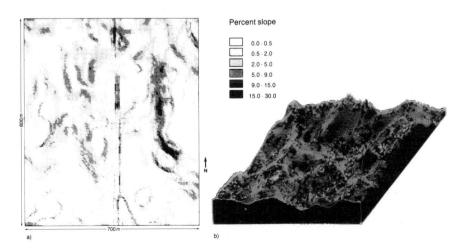

Figure 14.4. Plan (a) and three-dimensional (b) views of topography and slope gradient at the Lunty site.

remains until lost through infiltration or evaporation. At least 20 individual, unconnected depressions of significant areal extent can be recognized within the 700 m × 800 m study area.

Soils at the study site (Figure 14.5) strongly reflect the influence of water movement through the profile. The main soils are classified as Black Chernozem (Udic Hapliboroll and Argiboroll), Black Solonetz (Udic Natriboroll) and Humic Gleysol (Humaquent, Argiaquoll) according to the Canadian (Canadian Soil Survey Committee, 1978) and US (Soil Survey Staff, 1975) systems of soil classification. Chernozemic and solonetzic soils are located in the upper, convex portions of the landscape. They are predominantly well to moderately well drained, but possess a range of solonetzic features indicating some degree of past salination and possible contact with an elevated underlying water table. Soils in lower landscape positions and depressions are poorly to very poorly drained and are classified as Gleysols.

Almost all upland soils display some degree of solonetzic influence as expressed by weakly to strongly developed solonetzic B (natric) horizons. Soils higher in the landscape and closest to drainage divides are the least strongly solonetzic. Gleysolic soils display strongly developed eluvial horizons (Ae and Ahe) indicative of net downward movement of water within depressions. A significant feature, relative to studies of water movement, is the strong permeability contrast which may develop between highly permeable topsoil horizons (Ap, Ah and Ahe) and relatively impermeable B horizons (Bnt, Btnj and Bt).

An important practical consideration was that the Lunty site was already in use for other research work and was extensively instrumented to record the detailed meteorological and hydrological data required for this project. Existing activities provided automatic recording of temperature, rainfall, wind speed and direction and incoming solar radiation. Regular monitoring was also being carried out to record pan evaporation and to measure water table depths in over 150 piezometers. Snow pack surveys were carried out once a year, just before the beginning of the spring melt. The only required information not already recorded pertained to soil moisture

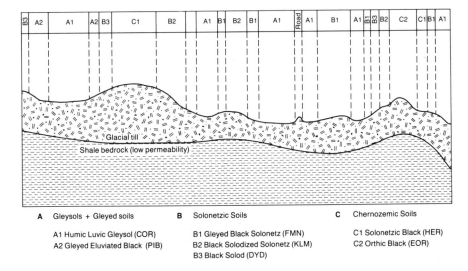

A Gleysols + Gleyed soils	**B** Solonetzic Soils	**C** Chernozemic Soils
A1 Humic Luvic Gleysol (COR)	B1 Gleyed Black Solonetz (FMN)	C1 Solonetzic Black (HER)
A2 Gleyed Eluviated Black (PIB)	B2 Black Solodized Solonetz (KLM)	C2 Orthic Black (EOR)
	B3 Black Solod (DYD)	

Figure 14.5. Schematic cross-section illustrating the relationship between topography and soils at the Lunty site.

content and moisture tension, saturated and unsaturated hydraulic conductivity, bulk density and porosity. Thus, only a minimum of additional instrumentation and field data collection was required for the site to meet all data requirements.

Characterization of the geometry of flow at the site

It was realized early in the investigation that attempting to produce detailed representations of the spatial and temporal patterns of run-off and ponding presented difficult technical problems. Given the highly dynamic behaviour of water flow and ponding, it was desirable to identify and base the modelling on relatively stable landscape features (i.e. flow paths and depressions).

Initially, research concentrated on defining fixed flow paths and, from them, the locations and characteristics of potential sites of 'depressional storage'. The concept of 'depressional storage' is reasonably well established and has been used as the basis for lumped conceptual models of farm scale flow in agricultural areas of Alberta (J. N. MacKenzie Engineering Ltd, 1986; W-E-R Engineering Ltd, 1987).

Once flow paths and depressional storage areas had been identified, it was possible to determine the geometry and topology of surface water flow into and between depressions. This geometry was used as the structural basis for the distributed hydrological model created to estimate the extent to which depressions are filled with water at any given time. The explicit definition of flow paths and storage and overflow characteristics of depressions permitted estimation of downslope storage potential and of possible downslope drainage impacts.

Determining the geometry of run-off flow and ponding permitted the following questions to be answered.

1. Where are the locations for depressional storage within the farm sized study area?
2. What is the total storage capacity of each depression?
3. How are the depressions connected to one another and what would be the sequence in which they would overflow into one another and contribute flow to downstream locations?
4. Are there any natural depressions with sufficient storage capacity to permit drainage of other, smaller, depressions and consolidation in a single on-farm location?

All the information required to determine where surface water can pond and to estimate the maximum extent and volume of 'depressional storage' was obtained by processing a detailed gridded digital elevation model (DEM). This study used a suite of interactive, PC based hydrological modelling utilities ('Watersh') (van Deursen and MacMillan, 1991) to define flow geometry and establish depressional storage capacity. The Watersh utilities are based on a simple raster representation of space and act on gridded elevation data. Numerous similar algorithms have been described by other authors including Mark (1983), Miller (1984), O'Callaghan and Mark (1984), Band (1985, 1986a, b), Jenson (1985), Jenson and Trautwein (1987), Martz and de Jong (1987), and Morris and Heerdegen (1988). The fundamental concepts and core algorithms used in Watersh are based principally on those given by Marks et al. (1984) with extensions similar to those proposed by Jenson and Dominique (1988).

In common with most similar programs, the Watersh utilities compute flow directions and flow paths, upstream element counts, catchment areas and stream

networks (see Figure 14.6). The initial algorithms were extended to enable account-
ing of flow into, through and out of depressions. The two principal modifications
required were:

1. the addition of a capability to determine the volume and location at which
   each depression overflowed and the neighbouring catchment and depression
   into which each overflowed; and
2. the provision of a capability to remove pits selectively based on pit size,
   area, maximum overspill volume and other criteria—this worked through a
   pit unblocking procedure that reversed flow directions from a given pit
   centre to its identified overspill location.

The gridded elevation data for the Lunty site were processed to define:

1. the location of all depressions and their associated catchments;
2. the maximum depth, areal extent and volume of each depression; and
3. the geometry controlling the interconnection of depressions and the
   sequence in which depressions were likely to overspill and connect.

Location of depressions and extent of catchments

Grid elements entirely surrounded by elements of higher elevation are identified by
the Watersh utilities as pit centres. The locations of all possible pit centres at the
Lunty site (Figure 14.7) were readily identified using this capability. Once the pit
locations were known, all grid elements draining to a given pit centre were identified
as belonging to a 'depressional catchment' associated with that pit. Initial processing identified numerous small catchment areas terminating in depressions of
minor area or volume (Figure 14.8(a)). Many of these small depressions were arte-
facts arising from the incomplete representation of the topographic surface afforded
by the gridded DEM. Others were of such minor extent and volume that they could
be ignored for the purposes of this study. The selective removal capability was used
to unblock pits of minor area or volume and define larger depressional catchments
draining into depressions of significant size and volume (Figure 14.8(b)).

The maximum depth, area and volume of depressions

The Watersh utilities were used to compute information on the extent and volume
of all significant depressions at the Lunty site. The algorithm used works by going
to each identified pit centre in turn and climbing up-slope to the next highest
adjoining grid element. This continues, one grid element at a time, until a grid
element is encountered that is not assigned to the current watershed, but drains into
a neighbouring watershed. As each cell is processed, the extent of the flooded area is
incremented by one grid cell and the volume of the pit is computed for all cells
flooded to the current fill level. The depression is considered to reach its maximum
extent and volume and to overflow when the flooded area reaches the first (lowest
elevation) cell draining into a different, neighbouring watershed area. This is the
outlet cell, or 'pour point'. The maximum volume and area computed for the depression up to this point is written to a watershed report file (Table 14.1). The Watersh
utilities were used to remove smaller pits once their attributes had been recorded so
that any larger, subsuming pits could also be identified and documented.

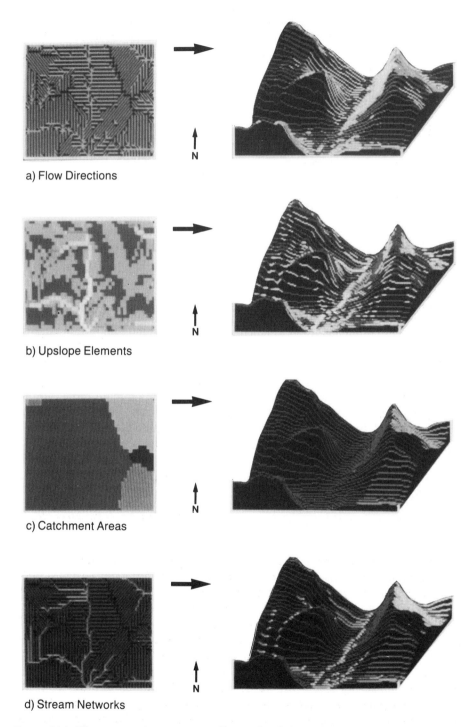

a) Flow Directions

b) Upslope Elements

c) Catchment Areas

d) Stream Networks

Figure 14.6. Illustration of output typically produced by raster based geomorphic run-off routing models.

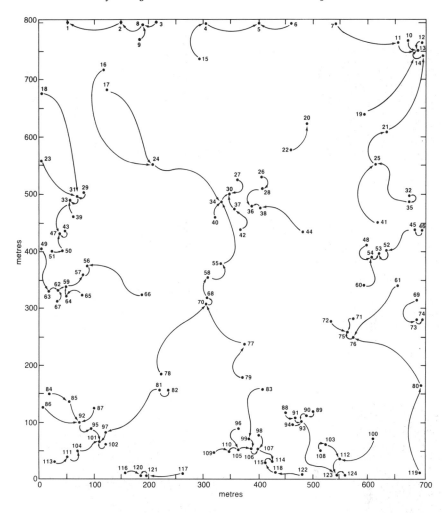

Figure 14.7. Location and estimated connectivity of all pit centres defined for the Lunty site.

Establishment of the geometry and sequencing of depressional connectivity

The information on pit volume, location, extent, depth and downslope connectivity computed by the Watersh utilities and output to the watershed report table (Table 14.1) was used to establish how each pit connected with and overspilled into its neighbour (Figure 14.7). It was also used to estimate the most likely sequence in which depressions would fill up, overspill and connect or coalesce.

The Watersh algorithms establish the geometry of connectivity by identifying the potential 'pour points' between each depression and all of its neighbouring watersheds. Water from any given depression will overspill through the lowest of the potential pour points to neighbouring watersheds, providing that overflow has not already occurred from the neighbour back into the depression in question. The lowest potential pour point for each depression is readily identified from Table 14.1. In cases where more than one potential pour point existed at the same elevation, it was necessary to direct overflow to only one. The model algorithm did not permit multiple overflow points as would happen in nature. Overflow was directed to the pour point leading to the lowest downslope pit in such cases. If several pits shared

(a)

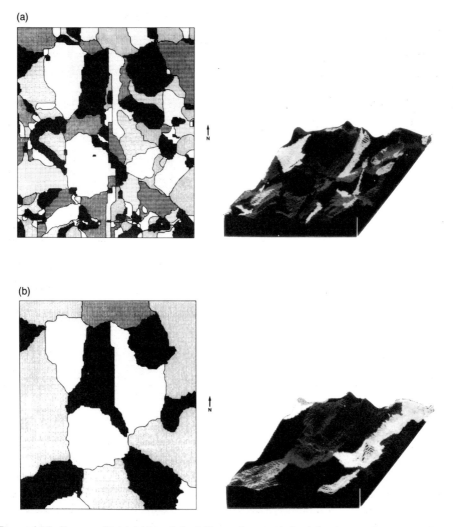

(b)

Figure 14.8. Extent of initial (a) and final (b) catchments defined for the Lunty site. Small pits have been removed in (b) but not in (a).

the same overspill elevation pointers were selected to be circular so as to ensure that all pits at that elevation overspilled into one another before overspilling into either lower, downslope pits or higher, subsuming pits.

The information contained in the watershed report table was also used to estimate the approximate sequence in which depressions would fill up and overspill. The ratio of a pit's volume to its catchment area (V/A ratio) provides an estimate of the amount of rainfall required to fall on the catchment before a pit will fill from empty and overspill. This was used to establish the likely sequence in which each pit would fill up and overspill. By working through this sequence, one pit at a time, and updating the watershed report table after each pit overspill, it was possible to estimate the likely sequence of pit connectivity (Figure 14.9).

Establishment of the actual sequence of connectivity was complicated by the existence of nested depressions (Figure 14.10). If, upon overflowing, the excess water

Table 14.1. *Example of the watershed report data produced by Watersh.*

Shed Num.	Area	Pit Row	Col.	Elev.	Vol.	Area	Drains to Pit	Out Row	Col.	Elev.	Over Row	Col.	Elev.	Pour Elev.
1	496	1	10	722.6	0.0	0	0	13	1	725.4	12	0	0.0	725.4
1	496	1	10	722.6	0.0	0	2	1	28	724.0	2	29	724.0	724.0
1	496	1	10	722.6	0.0	0	9	9	31	724.8	8	32	724.7	724.8
1	496	1	10	722.6	0.0	0	24	11	31	724.9	12	32	724.9	724.9
1	496	1	10	722.6	0.0	0	18	18	9	725.0	19	9	725.0	725.0
1	496	1	10	722.6	0.0	0	16	16	22	725.5	17	23	725.5	725.5
2	40	1	30	723.9	0'0	0	0	1	34	724.1	0	35	0.0	724.1
2	40	1	30	723.9	0.0	0	1	1	29	724.0	1	28	724.0	724.0
2	40	1	30	723.9	0.0	0	8	1	34	724.1	2	35	724.1	724.1
2	40	1	30	723.9	0.0	0	9	5	35	724.3	6	36	724.3	724.3
3	79	1	43	724.0	0.0	0	0	1	50	725.6	0	51	0.0	725.6
3	79	1	43	724.0	0.0	0	8	1	40	724.0	1	39	724.0	724.0
3	79	1	43	724.0	0.0	0	4	4	50	725.4	5	51	725.3	725.4
3	79	1	43	724.0	0.0	0	15	9	50	725.9	10	51	725.9	725.9
4	124	1	61	723.7	0.0	0	0	1	64	724.0	0	65	0.0	724.0
4	124	1	61	723.7	0.0	0	3	5	50	725.4	4	49	725.3	725.4
4	124	1	61	723.7	0.0	0	5	1	64	724.0	1	65	724.0	724.0
4	124	1	61	723.7	0.0	0	15	9	58	724.9	10	59	724.9	724.9
5	493	1	80	721.7	0.0	0	0	1	87	722.2	0	88	0.0	722.2
5	493	1	80	721.7	0.0	0	4	1	65	724.0	1	64	724.0	724.0
5	493	1	80	721.7	0.0	0	6	1	87	722.2	1	88	722.2	722.2
5	493	1	80	721.7	0.0	0	15	12	66	725.3	13	65	725.3	725.3
5	493	1	80	721.7	0.0	0	34	16	68	725.4	17	68	725.5	725.5
5	493	1	80	721.7	0.0	0	36	17	75	724.9	18	74	724.8	724.9
5	493	1	80	721.7	0.0	0	20	17	96	724.0	18	97	724.0	724.0

from an overflowing pit cascades into a separate, downslope depression (Figure 14.10(a)), there are no major problems. If, however, the water levels of the overflowing pit and its neighbour are identifical (Figure 14.10(b)), the pits coalesce instead of cascading from one into the other. In this situation, the two original pits form a new larger pit with unique catchment characteristics. After such occurrences, it was necessary to define updated catchment statistics for the new larger pit. In particular, it was necessary to locate a new outlet elevation and to update the maximum storage volume and area for the new depression.

This situation required the recognition of a hierarchy of depressions (Figure 14.9). Explicit linkages needed to be established between depressions. New depressions inherited some characteristics from their components (i.e. the catchment area of a new combined catchment is the sum of the areas of the two contributing catchments), but also had attributes that were uniquely their own (i.e. the new overspill location and maximum storage volume).

The automatic geomorphic feature extraction capabilities provided by the Watersh utilities permitted calculation of most of the data needed to assess the static features of depressional storage and depressional connectivity. We were able to determine where water can accumulate, the maximum amount of water that can accumulate and where water removed from one depression was likely to flow and reaccumulate. What the Watersh utilities did not provide, in this instance, was an estimate of the magnitude and timing of flow required to assess the likely extent and volume of depressions at any given time. For this, it was necessary to introduce a capability to model the physical processes influencing run-off and accumulation.

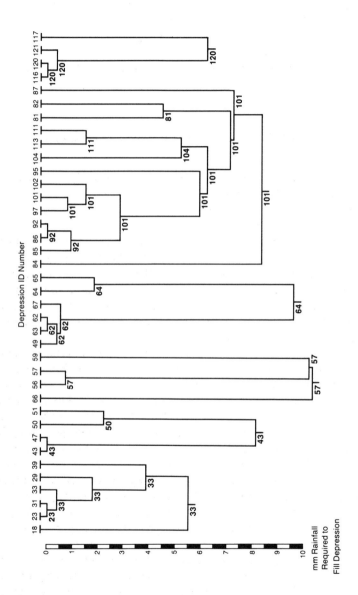

Figure 14.9. Dendogram illustrating the rainfall volume at which depressions are estimated to fill up and the sequence in which they are expected to overspill and interconnect (partial data set only).

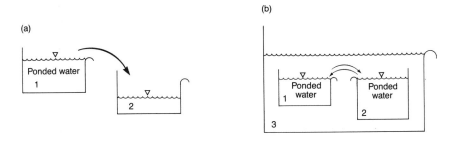

Figure 14.10. Illustration of the two principal conditions that may occur when a pit overspills.

Hydrological simulation modelling of magnitude and timing of run-off and accumulation

Hydrological simulation models are required to estimate the magnitude and timing of run-off so as to evaluate the likely extent and volume of ponded surface water at any given time. A brief description follows of the purpose built simulation models assembled to address this problem.

Physically based, distributed models are currently much in favour (Abbott *et al.*, 1986a, b; Anderson and Rogers, 1987). Such models attempt to represent process in a realistic and comprehensive manner through the use of finite difference solutions of the theoretical partial difference equations of mass, momentum and energy conservation (Bathurst, 1986a). They also represent space in the most highly disaggregated manner possible, given computational and memory constraints imposed by their complex representation of process.

An initial review of the literature suggested that a comprehensive physically based hydrological model such as the Système Hydrologique Européen (SHE) (Abbott *et al.*, 1986a, b) would be the most appropriate tool for simulating the timing and spatial pattern of run-off and ponding at a farm scale. Further investigation revealed serious drawbacks to this approach for the present problem. On the theoretical side, serious reservations about the use of complex physically based models for predictive purposes have been raised by several authors (Klêmes, 1978; Dooge, 1988; Beven, 1989). There is concern about the degree to which point field measurements (or worse, estimates) can be used to represent average conditions at the scale of model elements (Bathurst, 1986b; Beven, 1989). Similarly, concern has been expressed about whether equations based upon small scale soil–water physics can properly be applied to distributed elements that are usually many orders of magnitude larger than the volumes used to derive the equations (Dooge, 1988; Beven, 1989). In practical terms, effective use of highly distributed, physically based models has also been hampered by problems associated with obtaining and verifying the accuracy of data for very large numbers of model elements (Abbott *et al.*, 1986a; Bork and Rohdenburg, 1986).

Beven (1989) has suggested that use of highly complex physically based models like the SHE was appropriate when the primary aim was to study and understand process, but that such models were still too unreliable when the primary aim is prediction. Since the principal aim of the present research was to evaluate the predictive capacity of models that are available and feasible for farm scale analyses of water movement, an alternative to complex physically based models was required.

Support for a simpler approach is provided by Burt (1985) who noted that there was some merit in the development of less detailed pseudo-physical or conceptual models and Dooge (1988) who argued for the need to develop simplified models of flow behaviour at the meso scale at which most basin and catchment modelling occurs.

The approach adopted was to create two different purpose built models one semi-distributed (CATCHMOD) and the other fully distributed (DISTHMOD) to simulate the timing of flow and accumulation of water. Both approaches try to account for delays and losses in the delivery of rainfall or snow-melt to pit locations. Timing is related to overland travel distances and travel times from source locations to pit centres and to delay caused by water entering into storage in the soil along the way. The magnitude of water accumulating at pit centres is altered according to losses arising from evapotranspiration and infiltration both within the depressional ponds and along flow paths leading to the ponds.

A semi-distributed, object-oriented approach (CATCHMOD)

The catchment approach to topographically controlled hydrological modelling of depressions (CATCHMOD) approach was created to make use of the highly distributed representation of space afforded by a detailed DEM while at the same time minimizing the volume of data to be processed by defining a limited number of depressional catchment objects that can be modelled using a simplified lumped parameter approach. This approach provides the locational information needed to identify, in some detail, where water can accumulate but reduces spatial detail to permit simplified and rapid hydrological modelling (Figure 14.11).

Each depression, and its associated catchment, is defined as an object in the CATCHMOD approach. The factors that influence flow and ponding within each watershed were computed, using the Watersh utilities, and stored as attributes for each watershed object. The most significant information is the data on pit volume, catchment area and depressional connectivity discussed above. Watershed area affects the conversion of rainfall and snow-melt inputs to runoff. The total depressional storage capacity determines at what volume a given depression will overflow into its downslope neighbour. The downslope neighbour into which each depression overflows determines where water released through overflow will flow.

Other factors considered are losses within each catchment and delays in the arrival of run-off at the depression centre. Losses within each flooded depression are calculated directly. Evaporation from the free water surface is computed at a rate equal to the daily potential evaporation. Infiltration is allowed through the bottom of the pond at a rate equal to the measured saturated conductivity of the soil under the pond. Infiltration and evaporation at each pond are proportional to the computed area of the pond at a given volume.

Losses and delays in the portions of each catchment outside the ponds need to be estimated also. This portion of CATCHMOD has not yet been implemented. It is hoped to compute these indirectly according to the concept of contributing area (Beven and Kirkby, 1979; Beven *et al.*, 1984). This method is based on the geomorphological observation that saturated overland flow develops most rapidly and most frequently on low gradient slopes draining large upslope areas. This observation has been used by Beven and Wood (1983) to formulate an equation relating the likelihood of run-off generation at any given point to the ratio between the point's upslope area (α) and its slope gradient ($\tan \beta$). According to Beven and Wood (1983),

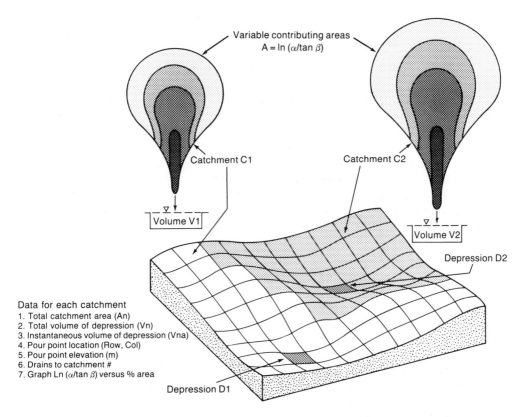

Data for each catchment
1. Total catchment area (An)
2. Total volume of depression (Vn)
3. Instantaneous volume of depression (Vna)
4. Pour point location (Row, Col)
5. Pour point elevation (m)
6. Drains to catchment #
7. Graph Ln (α/tan β) versus % area

Figure 14.11. Illustration of the concepts employed in the semi-distributed, variable source area approach used in CATCHMOD.

only those elements with values greater than a specified minimum for the parameter ln (α/tan β) will produce run-off for a specified antecedent moisture condition. Generalizing this concept to the catchment scale, the proportion of a catchment that will produce rapid run-off can be related to the relative extent of the catchment for which the value ln (α/tan β) exceeds a critical threshold. The data required to utilize the contributing area concept can easily be calculated using the Watersh utilities.

A first version of CATCHMOD is running with no loss and delay functions. This has produced estimates of pond location and volume (Figure 14.12) for a specified input of rainfall (10 mm). The estimates are very close to the observed distribution. This initial success provides encouragement that the models incorporating improved procedures for computing losses and delays will work.

A fully distributed cascading stores model (DISTHMOD)
The second approach developed for the study, distributed infiltration stores for topographically controlled hydrological modelling of depressions (DISTHMOD), maintains a fully distributed representation of space, whilst adopting a simplified conceptual approach to representing hydrological process (Figure 14.13). In it, losses and delays are computed explicitly for all grid elements using a modification of the model of conceptual stores proposed by Beven *et al.* (1984).

Figure 14.12. Three-dimensional view illustrating the location and volume of ponding simulated using the simplest implementation of CATCHMOD and a specified rainfall input of 10 mm.

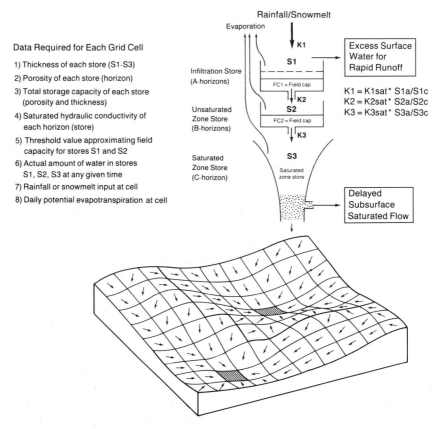

Figure 14.13. Illustration of the concept of cascading conceptual stores used in the fully distributed DISTHMOD approach.

Three conceptual stores are defined for each grid cell, corresponding roughly to a topsoil infiltration store (S1) a subsoil unsaturated zone store (S2) and a subsoil saturated store (S3). Infiltration is permitted into the infiltration store at a rate approximating the unsaturated hydraulic conductivity. This is estimated by reducing the value of saturated conductivity (K_{sat}) measured for the major horizons of all soils by the fraction S1a/S1c; where S1a is the actual amount of water in store S1 and S1c is the maximum total storage capacity of store S1. Infiltration losses are permitted from upper stores into lower stores (i.e. $S1 \rightarrow S2$ and $S2 \rightarrow S3$) at rates determined by the lowest value of unsaturated hydraulic conductivity between adjacent stores as limited by the moisture available in the contributing store or the space available in the receiving store. Movement into lower stores is not permitted if the actual amount of water in an overlying store is less than a critical threshold amount taken to approximate field capacity for that store.

Evaporation is computed initially at full potential from any free standing water (or run-off) on the soil surface and then, once there is no free water at the surface, from the upper store at a rate equal to the lesser of the remaining evapotranspiration potential or the unsaturated hydraulic conductivity of the soil. Once the water balance has been computed for any given cell, any water unable to infiltrate remains on the surface and is available for run-off into the next downslope grid cell.

The distributed modelling approach (DISTHMOD) uses data produced by the Watersh utilities to detemine the order in which grid cells are processed. Processing is by catchment area starting from the catchment area with the highest overspill elevation and working down to the lowest. Within each catchment, grid elements are processed along flow paths, in order, from the highest elevation to the lowest (i.e. the pit centre). In this way the correct order in which each grid cell delivers run-off to its downslope neighbour is maintained. In any given time step, the water balance for all upslope stores will have been computed and the excess water available for run-off determined before any given cell is processed.

An important difference between DISTHMOD and other distributed models occurs at the point when the last, or lowest, cell in any given catchment is processed. This cell must be either a pit centre or an edge cell draining to the 'outside world'. If it is a pit centre, a pond filling/emptying procedure is invoked. The procedure allows ponds to either: (a) fill, in which case they may either overspill or coalescence with adjacent ponds at the same elevation; or (b) empty, in which case they may bifurcate if the neighbour pond is at the same elevation, or break links with any lower, downslope catchment.

The pond change procedures make use of a relational database table containing certain useful 'pond statistics'. These include the volume of water currently in the pit, the maximum volumetric capacity of the pit, the identity of the watershed into which the pit will drain once full and the location of the 'pour point' at which water overspilling from the pond will enter the downstream catchment. The table also contains 'pointers' that indicate whether a pond is currently full and, if so, into which pit or watershed it is currently delivering its excess water. Using the pointers, it is possible to traverse through the pond data table to identify the location of the pond or pour point where excess run-off should be placed, or, in the case of net pond losses, the correct pond from which to reduce the recorded stored volume.

DISTHMOD permits explicit calculation of water balances at every grid location, but in a simplied manner that is feasible in an operational setting where both data availability and processing time are constraints. It is not and does not attempt to be, as comprehensive as other, more mathematically rigorous, physically based, distributed models but it may be more appropriate for the present problem. It does allow for a crude approximation to be made of likely losses and delays arising from evaporation and infiltration of run-off as it traverses flow paths to depressional centres. It is unique in its ability to account for flow into depressions and overflow or coalescence of ponds. Almost all other distributed models start with the assumption of completely integrated drainage and do not address 'depressional storage'.

Field mapping of actual water accumulation (ponding)

The actual locations and extent of surface water ponding were determined by detailed field surveys carried out for several periods in April and May of 1989 and 1990. Stakes were driven into the ground around the perimeter of all observable bodies of open surface water at weekly intervals beginning with the start of spring run-off and continuing until most of the non-permanent water had infiltrated or evaporated and only permanent ponds remained. The location and elevation of all perimeter stakes was accurately surveyed and digitized into a GIS. The field data have not yet been collated and processed. The digitized information will be used to determine the areal extent of the ponds. Volume will be determined by subtracting the observed elevation of ponding at each stage from the ground surface elevation recorded by the detailed DEM.

The field programme also entailed setting up instrumentation and conducting field experiments to collect data on the various soil properties required as input parameters for the process oriented modelling. The primary data recorded included soil moisture content and soil moisture tension, bulk density and porosity, and saturated hydraulic conductivity for the three major soil horizons representative of all significant soils found at the research site (see Table 14.2).

Comparing and evaluating simulation versus mapping results

No work has yet been carried out on comparing simulated versus mapped results. The intention is to determine for each time step, the degree to which simulated data agrees with the measured extent of water. This may take the form of simple comparisons of the percent of area where:

1. simulated estimates agree with observed distributions,
2. water is estimated to occur but is not actually observed, and
3. water is observed but is not estimated to occur.

Table 14.2. Field measurements and instrumentation at the Lunty site

Field measurements being undertaken	Instruments being used for measurement	Reference for the method
Saturated hydraulic conductivity (K_{sat})	Guelph permeameter	Reynolds and Elrick (1986)
Moisture content (% H_2O by volume)	Time domain reflectometer (Foundation Instruments)	Topp and Davis (1982), Topp (1987)
Moisture tension (bar)	Gypsum blocks (Soil Moisture Corp)	Tanner et al. (1948)
Bulk density and porosity	Giddings truck mounted core sampler	Macyk and Widtman (in preparation)

A slightly more quantitative approach might be to adopt the concept of 'model efficiency' often used to evaluate success in estimation of unit hydrographs (Ward, 1984; Bathurst, 1986a, b). The depth of water at every grid cell could be compared for estimated and measured distributions of surface water. Goodness of fit would be evaluated by calculating the sum of the squares of the deviations between observed and simulated distributions. This will yield a single statistic for model efficiency for any given time step. The individual time step efficiencies could be averaged to give an overall measure of model performance. This approach should allow for consistent and equitable comparison of the utility of different model approaches in providing the required data.

Summary

Considerable progress has been achieved to date on each of the major elements of this research (Table 14.3), but work still remains to complete model runs and compare model results with ground truth data. The most significant achievement to date has been in the area of geomorphic modelling. Geomorphic run-off routing models (Watersh) were obtained and modified to permit estimation of the location and extent of depressional storage capacity and to establish the pattern of connectivity between depressions at the Lunty site. The technique is expected to be applicable to any similar farm sized area.

Progress in the area of process-based hydrological modelling has resulted in the creation of a purpose-built, highly distributed hydrologic simulation model (DISTHMOD) appropriate for the present research. The control framework for a semi-distributed, lumped conceptual approach (CATCHMOD) has also been constructed, but elements relating to estimating timing and magnitude of run-off have not yet been implemented.

Table 14.3. Summary of research activities completed to date and remaining to be done.

Research activity	Present status	Future plans
Geomorphic characterization of Lunty site	Achieved using **Watersh** utilities	Illustrate using figures and dendograms
Hydrological modelling of magnitude and timing of flow at the Lunty site	Conceptual design completed for semi-distributed and distributed models CATCHMOD, DISTHMOD	Write computer programs to implement the conceptual designs
Field survey of actual water distribution and related soil characteristics	Pond extent mapped weekly from 1 April 1989 to 7 May 1989	Digitize pond extent, compute area and volume
	Field measurement of sat hydraulic conductivity (K_{sat}) moisture and tension	Distribute data over area and use as input for models
Comparison of simulated and observed distributions of surface ponding at *n* times	No progress to date (techniques for comparing reviewed)	Assemble data and compare results of simulated versus observed distributions

The actual distribution of ponded water at the Lunty site has been delineated for several time periods as a result of field work completed in April and May, 1989 and 1990. The data have been collated and initial review suggests that the measured distributions are not too far different from those estimated using even the simplest geomorphic modelling approach. Field measurements have been made of saturated hydraulic conductivity and monitoring of soil moisture content and soil moisture tension has been completed at several locations representing all major soils found at the site. These data are being used to establish initial conditions and to assign parameter values for the process based modelling.

GIS capabilities were found to be an essential enabling component of the research. Models for detailed simulation of spatial and temporal patterns of surface water flow are not included in most existing GIS. No existing GIS was available that was capable of producing the information on geometry and timing of water flow and accumulation required for the present study. GIS tools were, however, essential to collect the data required for modelling, to bring the data to a common format and spatial reference, and to integrate the data and models to facilitate model operation.

Conclusions

The aim of this research was to analyse a unique drainage situation which has not been addressed by previously developed models. Until now, models concerned with surface water hydrology have dealt with flow into and through depressions in one of three ways, namely:

1. they have ignored the issue of flow into depressions entirely;
2. they have recognized flow into depressions as a problem and have contrived to remove depressions from the data set as a preprocessing operation, thereby creating an artificial situation of fully integrated drainage; or
3. they have, at best, addressed it by including a 'black box' parameter for depressional storage whose value was most often determined by iterative calibration techniques.

A distributed model (DISTHMOD) has been explicitly designed to analyse flows into and out of depressions.

1. The model produces detailed spatial and temporal estimates of the varying regions contributing run-off flow into depressions and of the extent and volume of ponded water in depressions at any given time.
2. Depressions are allowed to fill and, once full, to overspill into the next downslope catchment or to be flooded by higher order, subsuming depressions.
3. The model permits explicit calculation of run-off according to the 'variable contributing area' concept of Beven and Kirkby (1979) through the application of a simplified water balance model at each grid cell and a sequential processing of grid cells from highest to lowest in any catchment.

A unique data set possessing very high spatial and temporal resolution for a large amount of accurately determined meteorological and soil/landscape data has been collected for a single farm scale study site. The data set and distributed model combine to permit:

1. estimation of run-off conditions at the farm scale;
2. detailed simulation modelling for a farm size project using detailed field scale data; and
3. estimation of depressional storage volume for an agricultural landscape and the development of initial area/depth/volume relationships.

The unusual nature of the problem of flow into depressions has required the creation of novel approaches and solutions. Three of the most important innovations are:

1. the DISTHMOD processing framework which processes grid cells sequentially by elevation within each catchment, thereby ensuring that a correct order is maintained for producing run-off from upslope cells before delivering it to downslope cells;
2. the use of a relational database structure for the model which permits the retention of depressions in the data set and the explicit calculation of flow into and out of depressions; and
3. the use of relational database tables to record attributes for depressions such as area or volume and the use of pointers in the tables to define the timing and topology of hydrological connectivity between depressions.

Results from the simplified model runs completed to data offer encouragement for the success and utility of the final models which are currently being run and will be reported on in a subsequent publication.

Acknowledgements

The authors would like to gratefully acknowledge the contribution made by Ir. Willem van Deursen, University of Utrecht, The Netherlands. The suite of interactive PC based programs for determining watershed geometry and topology (WATERSH) written by him was instrumental in realizing the objectives of the project. Enhancements and additions were made to these programs with his active assistance in order to provide the specialized data on depressions required for the research. Field work was conducted on private land belonging to Mr G. Lunty for whose cooperation the authors are grateful. Funding for the first author was provided by scholarships awarded by the Alberta Heritage Scholarship Fund and the UK Overseas Research Scholarship (ORS) Fund as well as by the Alberta Research Council.

References

Abbott, M. B., Bathurst, J. C., Cunge, J. A., O'Connel, P. E. and Rasmussen, J., 1986a, An introduction to the European Hydrological System—Système Hydrologique Européen, 'SHE', 1: History and philosophy of a physically based, distributed modelling system, *Journal of Hydrology*, **87**, 45–59.

Abbott, M. B., Bathurst, J. C., Cunge, J. A., O'Connel, P. E. and Rasmussen, J., 1986b, An introduction to the European Hydrological System—Système Hydrologique Européen, 'SHE', 2: Structure of a physically-based, distributed modelling system, *Journal of Hydrology*, **87**, 61–77.

Alberta Water Resources Commission, 1987, *Drainage Potential in Alberta: An Integrated Study. Summary Report*. Edmonton: Alberta Agriculture. Alberta Environment. Alberta Forestry, Lands and Wildlife.

Anderson, M. (Marv Anderson and Associates Ltd., Edmonton), 1987, *Drainage Potential in Alberta: An Integrated Study. Technical Report 6: Economics Component*. Edmonton: Alberta Environment. Alberta Agriculture. Alberta Forestry, Lands and Wildlife.

Anderson, M. G. and Rogers, C. C. M., 1987, Catchment scale distributed hydrological models: a discussion of research directions, *Progress in Physical Geography*, **11**, 28–51.

Band, L. E., 1985, Digitial elevation models and hydrologic information systems. Advanced technology for monitoring and processing global environmental data, *Proc. International Conference of the Remote Sensing Society and the Center for Earth Resources Management*, pp. 207–210. London: Remote Sensing Society/CERMA.

Band, L. E., 1986a, Analysis and representation of drainage basin structure with digital elevation data, *Proc. Second International Symposium on Spatial Data Handling*, pp. 437–450, Seattle, WA: International Geographical Union. Commission on Geographical Data Sensing and Processing and the International Cartographic Association.

Band, L. E., 1986b, Topographic partition of watersheds with digital elevation models, *Water Resources Research*, **22**, 15–24.

Bathurst, J. C., 1986a, Physically-based distributed modelling of an upland catchment using the Système Hydrologic Européen, *Journal of Hydrology*, **87**, 79–102.

Bathurst, J. C., 1986b, Sensitivity analysis of the Système Hydrologic Européen for an upland catchment, *Journal of Hydrology*, **87**, 103–123.

Beven, K. J., 1977, TOPMODEL—A physically based variable contributing area model of catchment hydrology, *Working Paper No. 183*, Leeds: School of Geography, University of Leeds.

Beven, K. J., 1989, Changing ideas in hydrology—the case of physically-based models, *Journal of Hydrology*, **105**, 157–172.

Beven, K. J. and Kirkby, M. J., 1979, A physically based, variable contributing area model of basin hydrology, *Hydrological Sciences Bulletin*, **24**(3), 43–69.

Beven, K. J. and Wood, E. F., 1983, Catchment morphology and the dynamics of runoff contributing areas, *Journal of Hydrology*, **65**, 139–158.

Beven, K. J., Kirkby, M. J., Scholfield, N. and Tagg, A. F., 1984, Testing a physically-based flood forecasting model (TOPMODEL) for three U.K. catchments, *Journal of Hydrology*, **69**, 119–142.

Bork, H.-R. and Rohdenburg, H., 1986, Transferrable parameterization methods for distributed hydrological and agroecological catchment models, *Catena*, **13**, 99–117.

Burt, T. P., 1985, Slopes and slope processes, *Progress in Physical Geography*, **9**, 582–599.

Canada Soil Survey Committee, 1978, *The Canadian System of Soil Classification, Publication 1646*, Research Branch, Canada Department of Agriculture.

Dooge, J. C., 1988, Hydrology in perspective, *Hydrological Sciences—Journal des Sciences Hydrologiques*, **33**, 61–85.

Heil, R. J., 1980, The digital terrain model as a data base for hydrological and geomorphological analysis, *Auto Carto IV. Proc. International Symposium on*

Cartography and Computing: Applications in Health and Environment, Vol. 2, pp. 132–139, Falls Church, VA: American Congress on Surveying and Mapping. American Society of Photogrammetry.

Jensen, N. E. and Wright, J. B. (Jensen Engineering Ltd., Olds), 1987, *Drainage Potential in Alberta: An Integrated Study. Technical Report 1: Drainage Engineering Component*. Edmonton: Alberta Environment. Alberta Agriculture. Alberta Forestry, Lands and Wildlife.

Jenson, S. K., 1985, Automated derivation of hydrologic basin characteristics from digital elevation model data. Auto Carto 7. Digital representations of spatial knowledge, in *Proc. Seventh International Symposium on Computer Assisted Cartography*, pp. 301–310, Washington, DC: American Society of Photogrammetry and the American Congress on Surveying and Mapping.

Jenson, S. K. and Dominique, J. O., 1988, Extracting topographic structure from digital elevation data for geographic information system analysis, *Photogrammetric Engineering and Remote Sensing*, **54**(11), 1593–1600.

Jenson, S. K. and Trautwein, C. M., 1987, Methods and applications in surface depression analysis, in Chrisman, N. R. (Ed.), *Auto Carto 8. Proc. Eighth International Symposium on Computer Assisted Carotgraphy*, pp. 137–144, Baltimore, MD: American Society for Photogrammetry and Remote Sensing and the American Congress on Surveying and Mapping.

J. N. MacKenzie Engineering Ltd, in association with Jensen Engineering Ltd, 1986. *Hydrologic Simulation: Olsen Project*, Edmonton: Drainage Branch, Alberta Agriculture.

Johnston, L. E., 1989, MAPHYD—a digital map-based hydrologic modelling system, *Photogrammetric Engineering and Remote Sensing*, **55**(6), 911–917.

Klêmes, V., 1978, Physically based stochastic hydrological analysis, *Advances in Hydroscience*, **11**, 285–356.

Koeln, G. T., Caldwell, P., Wesley, D. E. and Jacobson, J. E., 1986, *Inventory of Wetlands with Landsat's Thematic Mapper. Tenth Canadian Symposium on Remote Sensing*, 5–8 May 1986, pp. 153–162, Edmonton, Alberta, Canada. Ottawa: Canadian Society of Remote Sensing.

Koeln, G. T., Jacobson, J. E., Wesley, D. E. and Remple, R. A., 1988, Wetland inventories derived from Landsat data for waterfowl management planning, *Transactions of the North American Wildlife and Natural Resources Conference*, Vol. 53, pp. 303–310, Washington DC: Wildlife Management Institute.

Leskiw, L. A. (Pedology Consultants, Edmonton), 1987, *Drainage Potential in Alberta: An Integrated Study. Technical Report 3: Soils and Agronomy Component*. Edmonton: Alberta Environment. Alberta Agriculture. Alberta Forestry, Lands and Wildlife.

Macyk, T. M. and Whitman, Z., in preparation, *Measurement of Bulk Density in Reconstructed Soils in the Battle River area of Alberta*, Edmonton: Alberta Research Council, Terrain Sciences Department.

Mark, D. L., 1983, Automated detection of drainage networks from digital elevation models. Auto Carto Six. Automated cartography: International perspectives on achievements and changes, in Weller, B. (Ed.), *Proc. Sixth International Symposium on Automated Cartography*, Vol. 2, pp. 288–298. Baltimore, MD: American Society for Photogrammetry and Remote Sensing and the American Congress on Surveying and Mapping.

Marks, D., Dozier, J. and Frew, J., 1984, Automated basin delineation from digital elevation data, *Geo-Processing*, **2**, 299–311.

Martz, L. W. and DeJong, E., 1987, Using Cesium-137 to assess the variability of net soil erosion and its association with topography in a Canadian praire landscape, *Catena*, **14**, 439–451.

Miller, S. W., 1984, A spatial data structure for hydrologic applications, *Proc. International Symposium on Spatial Data Handling*, pp. 267–287, Geographisches Institut, Abteilung Kartographie/EDV, Universitat Zurich Irchelinterthurerstrasse 190, Zurich.

Morris, D. G. and Heerdegen, R. G., 1988, Automatically derived catchment boundaries and channel networks and their hydrological applications, *Geomorphology*, **1**, 131–141.

O'Callaghan, J. F. and Mark, D. M., 1984, The extraction of drainage networks from digital elevation data, *Computer Vision Graphics and Image Processing*, **28**, 323–344.

Reynolds, W. D. and Elrick, D. E., 1986, A method for simultaneous in-situ measurement in the vadose zone of field saturated hydraulic conductivity, sorptivity and the conductivity–pressure head relationship, *Ground Water Monitoring Review*, **6**(1), 84–95.

Rogers, C. C. M., Beven, K. J., Morris, E. M. and Anderson, M. G., 1985, Sensitivity analysis, calibration and predictive uncertainty of the Institute of Hydrology distributed model, *Journal of Hydrology*, **81**, 179–191.

Rohdenburg, H., Diekkuger, B. and Bork, H. R., 1986, Deterministic hydrological site and catchment models for the analysis of agroecosystems, *Catena*, **13**, 119–137.

Silfer, A. T., Kinn, G. J. and Hassett, J., 1987, A geographic information system utilizing the triangulated irregular network as a basis for hydrologic modelling, in Chrisman, N R. (Ed.), *Auto Carto 8. Proc. Eighth International Symposium on Computer Assisted Cartography*, pp. 129–136, Baltimore, MD: American Society for Photogrammetry and Remote Sensing and the American Congress on Surveying and Mapping.

Soil Survey Staff, 1975, Soil Taxonomy, a basic system for soil classification for making and interpreting soil surveys. Agriculture Handbook No. 436. Superintendent of Documents, Washington D.C.: Government Printing Office.

Stewart, R. E. and Kantrud, J. A., 1969, Proposed classification of potholes in the glaciated prairie region: Saskatoon Wetlands Seminar. Transactions of a seminar on small water areas in the prairie pothole region held February 20–22, 1967, *Canadian Wildlife Service Report Series No. 6*, pp. 57–69, Ottawa: Queens Printer, Department of Indian and Northern Development.

Tanner, C. B., Abrams, E. and Zubrinski, J. C., 1948, Gypsum moisture-block calibration based on electrical conductivity in distilled water, *Soil Science Society Proceedings*, **12**, 62–65.

Topp, G. C. and Davis, J. L., 1982, Measurement of soil water content using time domain reflectometry, *Canadian Hydrology Symposium: 82*, pp. 269–287, Ottawa: National Research Council of Canada.

Topp, G. C., 1987, The application of time-domain reflectometry (TDR) to soil water content measurement, *Proc. International Conference on Measurement of Soil and Plant Water Status*, Vol. 1, pp. 85–93, Utah State University, Logan, Utah.

University of Utrecht, 1987, Manual Modular Microcomputer Geographic Informaton System, PC-Geostat, University of Utrecht, Department of Geography, Heidelberglaan 2, P.O. Box 80.115, 3508 TC Utrecht, The Netherlands.

Van Deursen, W. P. A. and MacMillan, R. A., 1991, Watershed: general purpose utilities for hydrological and geomorphological processing of raster data. Internal publication, Department of Physical Geography, University of Utrecht, The Netherlands.

W-E-R Engineering Ltd, in association with J. N. MacKenzie Engineering Ltd. (Calgary), 1987, *Drainage Potential in Alberta: An Integrated Study. Technical Report 2: Hydrology/Hydraulics Components*, Edmonton: Alberta Environment. Alberta Agriculture, Alberta Forestry, Lands and Wildlife. (Includes appendices on the effect of precipitation rate and controlled versus uncontrolled drainage.)

Ward, R. C., 1984, Hypothesis testing by modelling catchment response, *Journal of Hydrology*, **67**, 281–305.

Weibel, R. and DeLotto, J. S., 1988, Automated terrain classification for GIS modelling, *Assessing the World: GIS/LIS 88, San Antonio, TX*, Baltimore, MD: American Congress of Surveying and Mapping. American Society of Photogrammetry and Remote Sensing.

Weibel, R., Heller, M., Herzog, A. and Brassel, K. E., 1988, Approaches to digital surface modelling, *First Latin American Conference on Computers in Geography*, San Jose, Costa Rica, pp. 143–163. San Jose, Costa Rica: University Estatal Distancia.

15

The effects of management on heath and mire hydrology: a framework for a geographic information system approach

A. M. Gurnell, P. J. Edwards and C. T. Hill

Introduction

This chapter considers the relationships between semi-natural vegetation communities, particularly dwarf shrub and mire communities, their management and hydrological processes. In considering these topics, the chapter addresses three important themes within current hydrological research: the monitoring and modelling of fundamental hydrological processes at the small-area scale; the requirement for input data to catchment-scale, physically based, distributed hydrological models; and the evaluation of the central role of vegetation in hydrology through its direct impact upon hydrological processes, its adjustment to hydrological processes and the role of its management in modifying and altering hydrological processes.

At a small-area (individual plant or small plot) scale, it is possible to measure and model the fundamental hydrological processes and process relationships, but over larger areas such an approach would be time-consuming and costly. Landscape management is best undertaken at the large-area (catchment) scale but, if we are to make use of knowledge derived from small area studies, surrogates are required to extrapolate key components of those studies in an efficient way. The use of surrogates is well illustrated by methods derived for assessing and managing catchment-scale soil erosion problems. The relationship between rainfall characteristics, soils, topography and vegetation type and cover can be readily evaluated for a specific site by instrumenting a small (field-sized) plot. There is great variability in soil loss from plot to plot, but by analysing a very large sample of plot-years, Wischmeier and Smith (1978) devised a method for estimating soil loss from field-sized plots—the universal soil loss equation (USLE). This equation contains indices of the various factors affecting soil loss at the field scale, which can be readily estimated using equations, tables and nomograms. Thus the USLE, or techniques based on the USLE indices, have been developed for catchment-scale soil conservation. For example, Dickinson *et al.* (1986) propose a model for small agricultural catchments based on the seasonal application of the USLE to fields within the basin and the translation of these field soil loss estimates into a sediment yield using a spatially variable delivery ratio based on soil, land use, flow length and the hydrological conditions of the flow path. A more physically based approach is described by de Roo *et al.* (1989), who develop a geographic information system (GIS) framework to apply ANSWERS (a distributed parameter model for estimating surface run-off and soil erosion, developed by Beasley and Huggins (1982)). The ANSWERS model employs distributed values of the USLE cropping (C), management (P) and erodibility (K) factors, but these are included with a large range of other hydrological and hydraulic parameters in physically based, component relationships rather than in the statistically derived relationship proposed by Wischmeier and Smith. Thus, the development of indices or surrogates, which represent processes or process controls and for which data are readily available or easily estimated over large areas, provides a means of extrapolating the fundamental understanding of processes at the small-area scale to a practical application of that understanding over larger areas.

Over the last 10 years, a number of catchment-scale, distributed hydrological models have been developed. These models have now reached the stage where they are capable of being applied and 'in view of their intrinsic complexity [they] necessitate a much fuller appreciation of model elements, algorithm alternatives and catchment representation than perhaps exponents and users of physically-based hydrological models have hitherto been accustomed to' (Anderson and Rogers, 1987). Whilst the properties of catchment-scale, distributed models are not relevant to the present discussion, the data requirements of such models are relevant. 'Although, the parameters represent measurable properties, such measurements may be at best very expensive and time-consuming to obtain and, at worst, so difficult to undertake as to be virtually impossible' (Anderson and Rogers, 1987). Any means of developing rapid, accurate representations of the detailed distribution of the required catchment properties will be an invaluable aid to developing applications of these models. Inevitably, such rapid means must use surrogates, and a particularly promising area is the use of remote sensing techniques, which have the ability to extrapolate many land surface properties of relevance to hydrology through the use of algorithms based upon the reflectance or emission of electromagnetic radiation of different wavelengths by different surfaces (e.g. Nicholson, 1989).

The role of vegetation in hydrology is particularly pertinent to the direct understanding and modelling of hydrological processes, to the development of surrogates for the spatial extrapolation of hydrological processes, and to the prediction of changes in hydrological processes as a result of catchment management. Vegetation has a wide range of direct effects upon hydrological processes, which have been illustrated by representative catchment, small plot and river channel studies (e.g. Clarke and Newson, 1978; Hino *et al.*, 1987; Dawson, 1989; Gregory and Gurnell, 1988) and have been represented to varying levels of complexity in hydrological models including, for example, the use of afforested percentage to estimate catchment water yields (Calder and Newson, 1979); the use of forest age/growth to estimate leaf area index, E_t/E_0 ratio (i.e. the ratio of actual evapotranspiration from a wet canopy as influenced by the leaf area index to potential evaporation estimated using a standard formula), canopy interception storage capacity, stress thresholds, proportion of roots in the topsoil horizon and thus to estimate the effects of forests on catchment water yields (Schulze and George, 1987); and the estimation of standing biomass, leaf area index, leaf fall and decomposition, and total vegetation crown cover, and their effects upon evapotranspiration loss, soil hydrological properties, overland flow and soil erosion (Kirkby and Neale, 1987). In addition, to the direct impact of vegetation on hydrological processes, semi-natural vegetation communities are closely adjusted to environmental controls, including the hydrological (particularly soil moisture) regime. For example, Walker (1985) describes the link between tundra vegetation and environmental gradients (including soil moisture, cryoterbation and temperature gradients) at three spatial scales in the Prudhoe Bay region of Alaska; Hupp and Osterkamp (1985), Hupp (1986) and Olson and Hupp (1986) relate vegetation to fluvial land-forms, soils and bedrock in catchments in Virginia, USA; and Gregory and Gurnell (1988) review the role of vegetation in identifying river channel changes. As a result, the vegetation communities can indicate, and thus provide a surrogate for, hydrological and fluvial processes and a basis for their extrapolation. Finally, vegetation cover is probably the most easily manipulated component of a drainage basin, and because of its direct influence on hydrological processes, numerous studies have illustrated the enormous impact of vegetation change on hydrology (e.g. Bosch and Hewlett, 1982; Trimble *et al.*, 1987). Vegetation management through such processes as cutting, burning or grazing can lead to changes in biomass and, in some circumstances, to immediate or gradual changes in the vegetation type. All such changes have hydrological implications.

This chapter will draw upon research undertaken by the authors, mainly in the New Forest, Hampshire, England, on the hydrological significance of dwarf shrub and mire communities. The discussion will centre on three main aspects of that research: the relationships between plant communities, their management and hydrological processes at the small-area scale; the extrapolation of the results of small-area studies over larger areas through the production of appropriate vegetation maps; the development of a GIS framework for catchment vegetation management.

The relationship between plant communities, their management and hydrological processes at the small-area scale

(i) The adjustment of vegetation to hydrological processes: dwarf shrub and mire vegetation and the soil moisture regime

In the context of semi-natural vegetation, plant communities are usually well adjusted to environmental gradients, including soil moisture gradients, and thus the species composition can often be related to the characteristics of the soil moisture regime. Gurnell (1981), showed that the composition of heath and mire vegetation communities on a hillslope in the New Forest, Hampshire were a good indicator of soil moisture conditions and, in particular, of the mean and variance of pressure head and water table levels in the underlying soil. Quantitative relationships between parameters of the water table regime and parameters of the vegetation composition were estimated and used to map zones of near-surface water tables under given antecedent conditions. The effectiveness of using vegetation composition to define runoff contributing areas over whole catchments was tested by further studies. Detrended correspondence analysis (DECORANA) (Hill and Gauch, 1980) was applied to information on the percentage cover of different plant species from over 500 quadrats distributed throughout the catchment (Gurnell *et al.*, 1985). This ordination technique identified a soil moisture related axis within the vegetation data and scores on this axis formed the basis for relating vegetation composition to soil water regime (observed over a large network of shallow wells distributed widely within the catchment) and for extrapolating areas of consistent soil water regime across the catchment. For example, Figure 15.1 illustrates water table exceedance probabilities associated with six classes of vegetation, ranging from mire vegetation (class 6) to dry heath vegetation (class 1) in relation to the baseflow generated from the monitored area of heath. Specifically, it was possible to map areas of particular

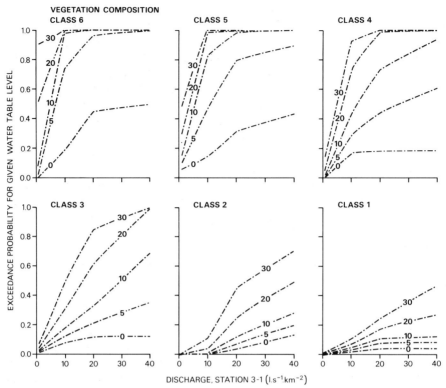

Figure 15.1. Water table level exceedance probabilities under six vegetation classes ranging from mire (class 6) to dry heath vegetation (class 1) in relation to baseflow discharge (l s⁻¹ km⁻²) for an area of heathland.

(Reproduced with permission from *Hydrological Processes*, 1987, **1**, 136.)

water table level exceedance probability (run-off contributing areas) from information on vegetation species composition (Gurnell and Gregory, 1986, 1987). Similar links between vegetation composition, soil water regime and, thus, run-off contributing area have been established by other researchers in relation to heath and mire communities. For example, Kent and Wathern (1980) noted the strong role of hydrological conditions in interpreting the results of a vegetation survey and classification for a small catchment on Dartmoor, England, and Prosser and Melville (1988) related six vegetation communities (including heath and sedge communities) to the underlying water table regime in a catchment near Sydney, Australia. Soil moisture characteristic curves were established in the laboratory using cores from soil derived from the area of the six plant communities. These curves illustrate that the unsaturated as well as the saturated soil water regimes are clearly reflected by the plant communities.

(ii) The adjustment of hydrological processes to vegetation: the example of snow storage

The impact of afforestation on hydrological processes such as evapotranspiration, interception loss and throughflow are well known, but all plant communities directly impact upon hydrological processes. For example, the influence of dry heath vegetation of varying age on hydrological processes will be discussed in section (iv). A good example of the way that vegetation communities directly influence hydrology is in the context of snow storage. In those areas of the world where winter precipitation occurs as snow, the distribution of snow depth at the onset of the spring snow melt is very important in influencing the shape and magnitude of the snow melt flood. Snow surveys during March 1985 in a small catchment in northern Finland illustrate the relationship between vegetation type and snow depth (Table 15.1). This relationship has long been recognised (Clark *et al.*, 1985).

Studies in Fennoscandia during the 19th century noted that wind desiccation of parts of plants protruding through the snow would lead to an adjustment of plant biomass to snow depth (Kihlman, 1890) and that there were contrasts in snow depth between forests and more open areas (Sundell, 1892). More recently, Yli-Vakkuri (1960) outlined the importance of vegetation for snow cover, ground freezing and timing of snow melt. Kallio *et al.* (1969) identified vegetation as the primary cause of local (catchment-scale) variations in snow depth and Karenlampi (1972) and Hiltunen (1980) illustrated how snow distribution is a function of wind interaction with topography and vegetation. All of these studies illustrate that a map of vegetation which reflects variations in standing biomass is likely to provide an excellent means of extrapolating limited snow depth measurements across large areas.

(iii) The adjustment of vegetation composition to vegetation management

Although, as outlined in (i) above, the species composition of semi-natural plant communities may provide a good indicator of hydrological processes, particularly of the soil water regime, management of that vegetation can result in adjustments of the hydrological regime, which may be sufficiently small, or may occur over sufficiently short a time, that there is no adjustment in the species composition of the vegetation. Alternatively, if the management persists and is associated with major changes in the hydrological (water quantity or quality) regime, a change in the species composition may result.

(iv) The adjustment of hydrological processes to vegetation management: the impact of burning or cutting of dry heath

In the New Forest, heathland, particularly dry heath, is managed by cutting and burning to improve the grazing (Tubbs, 1987; Webb, 1986). The impact of such

Table 15.1. Snow depth variation with vegetation categories Kilpisjärvi, 1985.

Vegetation category	Snow depth (cm)			
	mean (N = 50)	stand. devn.	range	coef. of var.
3. Medium altitude heath and rocks	5.4	4.6	0–17	85%
4. Low heath	10.0	5.9	0–31	59%
5. Shrub, meadow and marsh zone:				
Juniper & Heath	45.1	11.4	25–74	25%
Betula & Meadow	50.2	7.9	36–73	16%
6. Forest margin	85.8	12.5	63–126	15%
7. Birch forest:				
Upper slope	76.0	8.4	60–97	11%
Mid slope	69.0	8.4	50–90	12%
Lake margin	68.3	9.0	42–82	13%

Table 15.2. Summary of the hydrological characteristics of dry heath stands of different age (based on Gurnell et al., 1990).

Variable	Withybed	Beaulieu Road
Soil moisture content		wet antecedent conditions— no significant difference dry antecedent conditions— $4 > 3 > 2 > 1$
soil moisture tension		$1 > 2 > 3 > 4$
infiltration rate for 500 ml water (falling head infiltrometer)		$1 > 2 > 3 > 4$
stable infiltration rate after prewetting (constant head infiltrometer)		$1 > 2 > 3 > 4$
Interception capacity (canopy)		$4 > 3 > 2 > 1$
Interception capacity (litter)		$4 > 3 > 2 > 1$
Overland flow	$A > B$	

Dates of most recent burn: site A, 1979; B, 1965; 1, 1974; 2, 1970; 3, 1966; 4, 1962. Thus in 1980 the age of the heath on each site was: A, <1 year; B, 15 years; 1, 6 years; 2, 10 years; 3, 14 years; 4, 18 years.

management on hydrology is well illustrated by studies of the hydrological characteristics of stands of dry heath of different age (Gurnell *et al.*, 1990). The hydrological properties of dry heath stands were monitored during 1979 and 1980 at two sites in the New Forest; at one site (Beaulieu Road) four adjacent plots of dry heath, which were last burnt in 1974 (plot 1), 1970 (plot 2), 1966 (plot 3) and 1962 (plot 4) (i.e. plots of 6, 10, 14 and 18 years since burning in 1980) were studied, whereas at the second site (Withybed) two adjacent plots which were last burnt in 1979 (plot A) and in 1965 (plot B) (i.e. <1 and 15 years since burning in 1980), were studied. Full details of the field, laboratory and statistical techniques employed in the data collection and analysis of the studies are provided in Gurnell *et al.* (1990). Table 15.2 summarises the results of this research and shows that there were clearly identifiable differences in the hydrological properties of the different aged stands. The broader implications of the results summarised in Table 15.2 can be assessed in the context of the commonly occurring assemblage of heath and mire communities in the New Forest. Heath areas can be subdivided into dry heath (which usually occurs near the top of hillslopes), wet heath and mire areas (which usually occur downslope of the dry heath stands). This assemblage implies that the supply of soil moisture to wet heath and mire areas comes partly from the upslope dry heath areas, and that any change in the water relations of the dry heath (for example, as a result of management) may have an impact on the downslope soil water regime. In detail, it appears that runoff from managed dry heath may be partitioned differently between overland and throughflow, and that this, combined with the surface drainage dynamics of wet heath areas (Gurnell, 1978) could result in a change in soil water content and regime across the whole hillslope (Gurnell *et al.*, 1990).

Vegetation mapping as a basis for inferring hydrological processes

The examples cited above of the many links between vegetation and hydrological processes indicate that appropriately calibrated vegetation maps could provide a useful surrogate for certain types of hydrological map. Maps of vegetation cover

and species composition are more easily and rapidly compiled than maps of many
hydrological variables, but even so, field survey needs to be minimised if catchment-
scale maps are to be compiled and repeated readily. Traditional methods of vegetation mapping involve prolonged and detailed quadrat surveys, an analysis of the
quadrat-based data to establish vegetation classes, and then further field survey to
delimit the geographical boundaries of the vegetation classes. Visual interpretation
of air photographs or standard false-colour composites of digital data from airborne
or satellite platforms, provide a more rapid means of defining the boundaries of the
vegetation classes, once the classes have been established using ground survey.
Gurnell and Gregory (1987) describe the contrasting results of undertaking interpretation of air photographs and false colour composites, derived from simulated
SPOT HRV data, to map soil moisture-related vegetation classes defined by
detrended correspondence analysis of a quadrat survey for a catchment in the New
Forest (Figure 15.2).

Remote sensing techniques offer enormous potential for producing
hydrologically-relevant vegetation maps because of the availability of data for different dates and of variable spatial and spectral resolution (Price, 1986). The
opportunities for the development of algorithms to enhance hydrologically-relevant
properties of these data so that visual interpretation of boundaries can be more

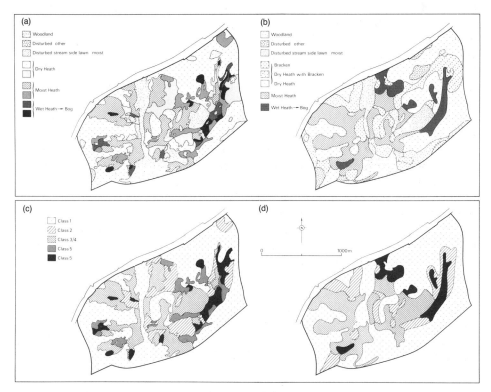

*Figure 15.2. (a) Vegetation composition classification based on detailed field survey and air
photograph interpretation. (b) Vegetation composition classification for the same area as map (a)
based on interpretation of a false-colour composite derived from 20 and 10 m resolution SPOT
HRV data. (c) Winter rainfall acceptance potential map derived from the vegetation classification presented in (a). (d) Winter rainfall acceptance potential map derived from the vegetation
composition classification presented in (b).*
(Reproduced with permission from *Hydrological Processes*, 1987, **1**, 139.)

effective, or to directly classify the data to produce vegetation-based hydrological maps are considerable. Milton *et al.* (1986), for example, describe the impact of changing spatial resolution on the visually identifiable vegetation classes by resampling and inspecting standard false colour composites. They note the differentiation of heath, mire, grassland and woodland at 30 m resolution; the additional distinction between wet heath and mire at 20 m resolution; and the discrimination of areas of disturbance within the heath and mire areas at 10 m resolution. They then consider the additional information that can be revealed to aid visual interpretation by combining different spectral bands using a variety of algorithms applied to both 20 m (three band) and 10 m (panchromatic) resolution simulated SPOT HRV data. The two most useful composites were found to be a modulated false-colour infra-red composite (blue gun = SQRT (panchromatic × green); green gun = SQRT (panchromatic × red); red gun = (0.75 × infra-red) + (0.25 × panchromatic)) and a visible band composite produced by combining the green band of the 20 m data (resampled to 10 m pixels) on the blue gun of the monitor, with the red band (also resampled) on the green gun, and the panchromatic band on the red gun. The first image acted as a mask to separate woodland and grassland from heathland and the second image allowed more subtle variations within the heathland and mires to be visually interpreted and mapped.

An automated classification of vegetation from remotely-sensed data would provide an even more rapid means of producing maps for large areas. However, the success of such an approach lies in finding the appropriate spatial resolution and combination of spectral bands to discriminate between the classes of interest. Clark *et al.* (1985) describe a simple means of analysing Landsat MSS (Multispectral Scanner) and TM (Thematic Mapper) data to provide a vegetation map for northern Finland that is indicative of potential snow storage. Analysis of ground radiometric and vegetation survey data indicated that a satisfactory discrimination of the vegetation classes that were related to differences in snow storage (i.e. the classes in Table 15.1) could be achieved by a simple density slice of an MSS band 7 : 5 ratio. The resulting vegetation map could be used to extrapolate snow storage immediately prior to the melt season by calibrating it with a small sample of snow depth values from the various vegetation categories. Wardley *et al.* (1987) demonstrate from an analysis of ATM data (collected from an altitude of 2000 m to give a nominal resolution of 5 m, for 11 bands) and ground data (collected using a Milton Multiband Radiometer to represent TM bands 2, 3 and 4 and a fourth band with peak sensitivity at 1.5 μm, and measured to represent spatial variability in vertical reflectance and the spectral reflectance indicatrix of the vegetation canopies) that 'there is sufficient variability between a variety of heathland vegetation types to warrant some type of explanatory model. Not only is there variability between sites of dissimilar vegetation type, but also more subtle variations between similar vegetation types at different stages of growth'. These conclusions are of great significance if we are to attempt to analyse remotely sensed data to derive vegetation and hydrological maps. Discrimination between different types of vegetation will permit the identification of areas of different soil moisture and overland flow regime, and discrimination within classes will allow us to take account of vegetation management in developing maps relevant to the spatial pattern of hydrological processes. The additional potential of SPOT data for identifying three-dimensional properties of the land and vegetation surface relevant to hydrology (watersheds, topographic characteristics, vegetation biomass) has yet to be evaluated in the context of drainage basins containing heath and mire communities.

Gurnell and Gregory (1987) take the use of vegetation maps as surrogates of hydrological processes a stage further by translating the vegetation classes interpreted from remotely sensed data into classes of winter rainfall acceptance potential (WRAP) (Farquharson *et al.*, 1978) using quantitative links between vegetation species composition, water table level exceedance probabilities (see Figure 15.1) and soil permeability within the heath and mire areas of the New Forest. This, like the snow storage map of Clark *et al.* (1985), is a true hydrological map derived using vegetation properties as surrogate hydrological indices.

The GIS approach

Vegetation patterns derived from remotely sensed sources and calibrated to represent the spatial distribution of hydrological stores, processes and indices, can be combined with digital terrain information, to provide the bases for developing a GIS that could evaluate seasonal patterns of certain hydrological variables, response patterns to particular rainfall events and vegetation management scenarios, and that could provide a means of storing and displaying inputs to and outputs from distributed hydrological models, so that the hydrological response to sequences of hydrological events can be evaluated.

The critical role of topography, including extremely small topographic features, in influencing hillslope hydrology has been widely reported (e.g. Anderson and Burt, 1978; Beven and Kirkby, 1979; Anderson and Kneale, 1982; Burt and Butcher, 1985). Algorithms to derive topographic landscape units that are appropriate to hydrological modelling from digital elevation data (spot heights and contours) have also received much research attention (e.g. Zevenbergen and Thorne, 1987; Moore *et al.*, 1988), permitting the derivation of slope angle, curvature and aspect classes and, in the case of Moore *et al.*, also upslope contributing area classes, all based on irregular-shaped grid units whose boundaries correlate with downslope and cross-slope trajectories. Moore *et al.*, go on to show how the topographic information associated with the grid units can be used to predict the natural occurrence of saturated zones and the patterns of potential daily solar radiation receipt across the catchment. Band (1989) defines drainage basin structure (i.e. watershed, sub-catchments and hillslope segments and the relationship to the drainage network, through thresholds of drainage area required to support a river channel) from digital elevation data, and goes on to use the resulting data to simulate runoff response using a distributed hydrological model (TOPMODEL) (Beven and Wood, 1983) assuming constant soil and vegetation cover.

The research reviewed in this chapter provides a suite of additional hydrologically relevant relationships that have been extrapolated from semi-natural vegetation patterns in addition to those that can be readily estimated from digital elevation data. Thus, in the context of the semi-natural plant communities discussed in this chapter, the hydrological information content of the digital elevation data could be greatly enhanced by the addition of hydrologically calibrated vegetation maps in a GIS. Such a system could be used to superimpose surface roughness, interception loss, soil water regime, infiltration regime and snow storage distributions onto the topographic pattern. Such a combination of topographic and vegetation-related information provides a firm basis for applying algorithms within the GIS to estimate the hydrological condition of the catchment during different

seasons of the year (combining the inferred soil moisture and infiltration characteristics from field-calibrated exceedance probability estimates and the interception
loss resulting from field-calibrated, seasonally variable plant biomass). The GIS
would also be capable of examining scenarios associated with different types and
locations of vegetation management (estimating the at-a-site and catchment-wide
hydrological impact) and the occurrence of specific rainfall events associated with
predefined antecedent hydrological conditions. For example, GIS would allow conclusions to be derived on the hydrological consequences of changes in grazing pressure. To do this it is necessary to establish the relationship between plant biomass
and animal use for various habitat types. A prediction of how changes in total
numbers of grazing animals might affect their distribution over the area would be
needed, and could be based upon the known preference of animals for particular
communities. Finally, the GIS can be designed, through the use of user exits, to act
as a shell to pass data to distributed hydrological models and to receive and display
the catchment response defined by such models to particular sequences of hydrological events.

References

Anderson, M. G. and Burt, T. P., 1978, The role of topography in controlling
    throughflow generation, *Earth Surface Processes*, **3**, 331–344.
Anderson, M. G. and Kneale, P. E., 1982, The influence of low-angled topography
    on hillslope soil-water convergence and stream discharge, *Journal of Hydrology*,
    57, 65–80.
Anderson, M. G. and Rogers, C. C. M., 1987, Catchment scale distributed hydro-
    logical models: a discussion of research directions, *Progress in Physical Geog-
    raphy*, **11**, 28–51.
Band, L. E., 1989, A terrain-based watershed information system, *Hydrological Pro-
    cesses*, **3**, 151–162.
Beasley, D. B. and Huggins, L. F., 1982, *ANSWERS—User's Manual*, West Lafay-
    ette: Department of Agricultural Engineering, Purdue University.
Beven, K. J. and Kirkby, M. J., 1979, A physically-based, variable contributing area
    model of basin hydrology, *Hydrological Sciences Bulletin*, **24**, 43–69.
Beven, K. J. and Wood, E. F., 1983, Catchment geomorphology and the dynamics of
    runoff contributing areas, *Journal of Hydrology*, **65**, 139–158.
Bosch, J. M. and Hewlett, J. D., 1982, A review of catchment experiments to deter-
    mine the effect of vegetation changes on water yield and evapotranspiration,
    Journal of Hydrology, **55**, 3–23.
Burt, T. P. and Butcher, D. P., 1985, Topographic controls of soil moisture distribu-
    tions, *Journal of Soil Science*, **36**, 469–486.
Calder, I. R. and Newson, M. D., 1979, Land-use and upland water resources in
    Britain—a strategic look, *Water Resources Bulletin*, **15**.
Clark, M. J., Gurnell, A. M., Milton, E. J., Seppala, M. and Kyostila, M., 1985,
    Remotely-sensed vegetation classification as a snow depth indicator for hydro-
    logical analysis in sub-arctic Finland, *Fennia*, **163**, 195–216.
Clarke, R. T. and Newson, M. D., 1978, Some detailed water balance studies of
    research catchments, *Proceedings of the Royal Society of London Series A*, **363**,
    21–42.

Dawson, F. H., 1989, Ecology and management of water plants in lowland streams, *Reports of the Freshwater Biological Association*, **57**, 43–60.

Dickinson, W. T., Rudra, R. P. and Wall, G. J., 1986, Identification of soil erosion and fluvial sediment problems, *Hydrological Processes*, **1**, 111–124.

Farquharson, F. A. K., Mackney, D., Newson, M. D. and Thomasson, A. J., 1978, Estimation of runoff potential of river catchments from soil surveys, *Soil Survey Special Survey No. 11*, Harpenden, Herts: Rothampstead Experimental Station.

Gregory, K. J. and Gurnell, A. M., 1988, Vegetation and river channel form and process, in Viles, H. A. (Ed.), *Biogeomorphology*, pp. 11–42, Oxford: Basil Blackwell.

Gurnell, A. M., 1978, The dynamics of a drainage network, *Nordic Hydrology*, **9**, 293–306.

Gurnell, A. M., 1981, Heathland vegetation, soil moisture and dynamic contributing area, *Earth Surface Processes and Landforms*, **6**, 553–570.

Gurnell, A. M. and Gregory, K. J., 1986, Water table level and contributing area: the generation of runoff in a heathland catchment, in *Conjunctive Water Use (Proc. Budapest Symposium, July 1986), IAHS Publication No. 156*, pp. 87–95, Wallingford: IAHS Press.

Gurnell, A. M. and Gregory, K. J., 1987, Vegetation characteristics and the prediction of runoff: analysis of an experiment in the New Forest, Hampshire, *Hydrological Processes*, **1**, 125–142.

Gurnell, A. M., Gregory, K. J., Hollis, S. and Hill, C. T., 1985, Detrended correspondence analysis of heathland vegetation: the identification of runoff contributing areas, *Earth Surface Processes and Landforms*, **10**, 343–351.

Gurnell, A. M., Hughes, P. A. and Edwards, P. J., 1990, The hydrological implications of heath vegetation composition and management in the New Forest, Hampshire, England, in Thornes, J. B. (Ed.), *Vegetation and Erosion*, pp. 179–198, Chichester: Wiley.

Hill, M. O. and Gauch, H. G., 1980, Detrended correspondence analysis: an improved ordination technique, *Vegetatio*, **42**, 47–58.

Hiltunen, R., 1980, Tuulenpiekseman ja lumenviipyman lampotila- ja lumioloista Pikku-Mallalla. (Temperature and snow conditions on snow beds and wind-exposed places on Pikku-Malla), *Luonnon Tutikija*, **84**, 11–14.

Hino, M., Fujita, K. and Shutto, H., 1987, A laboratory experiment on the role of grass for infiltration and runoff processes, *Journal of Hydrology*, **90**, 303–325.

Hupp, C. R., 1986, The headward extent of fluvial landforms and associated vegetation on Massanutten Mountain, Virginia, *Earth Surface Processes and Landforms*, **11**, 545–556.

Hupp, C. R. and Osterkamp, W. R., 1985, Bottomland vegetation distribution along Passage Creek, Virginia in relation to fluvial landforms, *Ecology*, **66**, 670–681.

Kallio, P., Laine, U. and Makinen, Y., 1969, Vascular flora of Inari Lapland: 1. Introduction and lycopodiaceae–polypodiaceae, *Reports from the Kevo Subarctic Research Station*, **5**, 1–108.

Karenlampi, L., 1972, Comparisons between the microclimates of the Kevo ecosystem study sites and the Kevo meteorological station, *Reports from the Kevo Subarctic Research Station*, **9**, 50–65.

Kent, M. and Wathern, P., 1980, The vegetation of a Dartmore catchment, *Vegetatio*, **43**, 163–172.

Kihlman, O., 1890, Pflanzenbiologische studien aus Russisch Lappland. Ein beitrag zu kenntnis der regionalen gleiderung an der polaren waldgrenze, *Acta Societatis Fauna et Flora Fennica*, **6**, 11–263.

Kirkby, M. J. and Neale, R. H., 1987, A soil erosion model incorporating seasonal factors, in Gardiner, V. (Ed.), *International Geomorphology*, Part II, pp. 189–210, Chichester: Wiley.

Milton, E. J., Gurnell, A. M. and Clark, M. J., 1986, Heathland vegetation and hydrological mapping from simulated SPOT HRV data, in *Heathland Vegetation Survey and Hydrological Mapping from Simulated SPOT HRV Data*. (Final report to NRSC on research conducted as part of the 1984 SPOT HRV simulation campaign.)

Moore, I. D., O'Loughlin, E. M. and Burch, G. J., 1988, A contour-based topographic model for hydrological and ecological applications, *Earth Surface Processes and Landforms*, **13**, 305–320.

Nicholson, S.E., 1989, Remote sensing of land surface parameters of relevance to climate studies, *Progress in Physical Geography*, **13**, 1–12.

Olson, C. G. and Hupp, C. R., 1986, Coincidence and spatial variability of geology, soils and vegetation, Mill Run watershed, Virginia, *Earth Surface Processes and Landforms*, **11**, 619–629.

Price, M., 1986, The analysis of vegetation change by remote sensing, *Progress in Physical Geography*, **10**, 473–491.

Prosser, I. P. and Melville, M. D., 1988, Vegetation communities and the empty pore space of soils as indicators of catchment hydrology, *Catena*, **15**, 393–405.

de Roo, A. P. J., Hazelhoff, L. and Burrough, P. A., 1989, Soil erosion modelling using 'answers' and geographical information systems, *Earth Surface Processes and Landforms*, **14**, 517–532.

Schulze, R. E. and George, W. J., 1987, A dynamic, process-based, user-oriented model of forest effects on water yield, *Hydrological Processes*, **1**, 292–307.

Sundell, A. F., 1892, Snotackets hojd i Finland Januari–Maj 1891, *Fennia* **7**, 50–51.

Trimble, S. W., Weirich, F. H. and Hoag, B. L., 1987, Reforestation and the reducton of water yield on the Southern Piedmont since circa 1940, *Water Resources Research*, **23**, 425–437.

Tubbs, C., 1987, *The New Forest: A Natural History*, London: Collins, New Naturalist Series.

Walker, D. A., 1985, Vegetation and environmental gradients of the Prudhoe Bay region, Alaska. *CRREL Report 85-14*, Hanover, New Hampshire: US Army Cold Regions Research and Engineering Laboratory.

Wardley, N. W., Milton, E. J. and Hill, C. T., 1987, Remote sensing of structurally complex semi-natural vegetation—an example from heathland, *International Journal of Remote Sensing*, **8**, 31–42.

Webb, N., 1986, *Heathland*, London: Collins, New Naturalist Series.

Wischmeier, W. H. and Smith, D. D., 1978, Predicting rainfall erosion losses—a guide to conservation planning, *USDA, Science and Education Administration Handbook No. 537*, USDA.

Yli-Vakkuri, P., 1960, Metsikoiden routa-jalumisuhteista. Summary: snow and frozen soil conditions in the forest, *Acta Forestalia Fennica*, **71**(5).

Zevenbergen, L. W. and Thorne, C. R., 1987, Quantitative analysis of land surface topography, *Earth Surface Processes and Landforms*, **12**, 47–56.

16

Use of geographic information systems for interpreting land-use policy and modelling effects of land-use change

R. Aspinall

Geographic information systems—a role in land-use decision-making at a policy level

Much of the current discussion on geographic information systems (GIS) has a pre-occupation with technical aspects; design, specification, hardware and software requirements, methods for data input, and data storage structures all feature in the literature (Rhind, 1988). Thus far however, there has been relatively little material published on applications of GIS methods, or possible roles for GIS-based methodologies in science or society. Such a deficiency exists in the UK, despite the conclusions of the Chorley Report (Department of the Environment, 1987; Chorley, 1988).

An area in which GIS technology and associated methodologies have the potential to make an important impact is land-use planning, and specifically in investigating the possible effects of implementation of policy (Worrall, 1989). Although policy statements represent a necessary early stage in attempts to solve problems, recent evaluation of planning processes and implementation procedures has shown the complex background against which planners attempt to convert policy into action. As a result, policy and its implementation are often rather less than effective in providing solutions to actual problems (Cloke, 1987).

This chapter presents a framework for applying GIS based methods to a wide range of land management and land-use planning issues including the interpretation and formulation of land-use policy. The framework allocates a central role to GIS in integrating policy with land resources and land use, thereby providing a powerful tool for land managers, planners, and policy-makers. A modelling approach to land use is necessarily included, models allowing interpretation of the likely effects of policy statements by attempting to predict the extent to which the policies concerned will solve the problems they are intended to address. In this sense, the framework is intended to contribute to application of GIS in land use as proposed by Chorley:

> GIS as a tool is about aiding managers to carry out their jobs more efficiently and effectively, and, more particularly, about better decision-making'. (Chorley, 1988: 3)

A GIS-based framework

The conceptual framework for applying GIS based method to land-use policy and change is illustrated in Figure 16.1. It is intended to provide a means of evaluating a plan, policy, or directive in terms of both its direct and indirect consequences on the biophysical and socio-economic environments. Within constraints of current models this allows plans, policies and directives to be tailored to produce desired effects and is a contribution to decision-making processes and policy *formulation* rather than merely a system for policy *interpretation*.

Land-use policy can be interpreted within GIS using a modelling approach (Jeffers, 1980). Output in the form of maps showing areas in which land-use changes are more likely to occur, and statistics, graphs and tables summarizing this information according to a variety of specified spatial units (for example, by agricultural parish) provides a direct feedback to policy. Such output allows land-use implications to be discussed. This level of interpretation and critical examination of likely consequences of policy will, in itself, significantly improve the possibility of obtaining desired land-use outcomes. The predicted land-use changes can also form input for GIS-based impact assessment. This route provides a means of establishing indirect consequences of policy implementation. Both environmental and socio-economic impact assessments are possible within GIS using a modelling approach (Openshaw *et al.*, 1987; Walker and Moore, 1988). As for the output from land-use modelling, impact-assessment output may be in map, graph, statistics or table form, GIS overlay facilities allowing easy summary of information according to specified spatial units of interest to different groups or users.

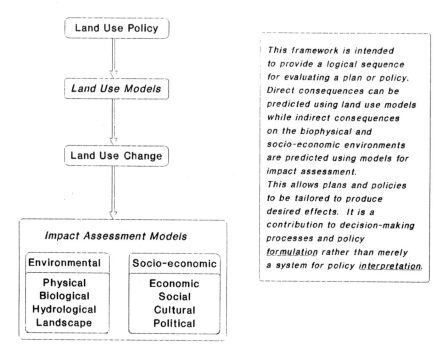

Figure 16.1. A conceptual framework for applying GIS-based methods to land-use policy and land-use change.

Much of the current power of GIS lies in the use of overlay techniques for the synthesis of varied data layers to produce new information. Requirements of models, especially those for impact assessment, highlight an additional use of the overlay facility. Impact assessments and models may require information on spatial bases other than that in which data are stored or pre-analysed; overlay can be used to *restructure* output, such that all inputs to subsequent analyses are on a common basis. This is particularly useful for relating predicted land-use changes to statistical information such as the parish agricultural statistics, or to land management units, such as catchments (Newson, 1987) or farm–estate boundaries.

Models and GIS

Much present modelling activity based on GIS makes extensive use of overlay, although this has limited application if operated solely as a process of identifying areas where particular combinations of conditions exist within the basic datalayers of a geographic database (Burrough, 1986). Prediction of the direct and indirect consequences of a policy draws on a variety of modelling approaches capable of incorporation into GIS.

Four main types of model are available: rule-based, knowledge-based, inductive-spatial, and geographic (Figure 16.2).

Rule-based modelling linked to GIS makes extensive use of overlay and includes sieve mapping which is already widely used in planning (see, for example, Strathclyde Regional Council 1988). Rule-based models can be modified within GIS to allow use of weightings, both for different datalayers being overlaid and for the different features represented within a datalayer (Quarmby *et al.*, 1988). This has

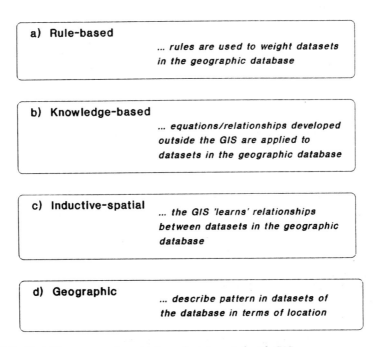

Figure 16.2. Modelling approaches which can be integrated with GIS.

application in land use planning, allowing land to be zoned according to the relative importance of different data. The allocation of weights allows the rules whereby decisions are made to be expressed formally and this increases opportunity for discussion in evaluating land use outcomes, the rule-base and resultant output scenario both being subject to scrutiny. Models using rule-based weighting of data can be developed for many applications from experience in land evaluation (Davidson, 1986; Bibby *et al.*, 1982; Bibby *et al.*, 1988) and from results of scientific experiment. Alternatively existing planning and development constraints from a variety of sources can be integrated within the system.

Apart from models employing weighting for rule combination, development of models employing data *transformation* is required (Newson, 1988). It is here that knowledge-based models have potential for great impact. This approach to modelling within GIS uses established equations and relationships to interpret datasets (e.g., Sivertun *et al.*, 1988). Frequently the knowledge-base has been developed from detailed scientific study but the applicability of exisiting models to spatial data needs testing and further models which are specifically *spatial* require development. Areas of land use where there is potential for application of knowledge-based models include pollution prediction, hydrological forecasting and aspects of ecological land management. Additionally these models require the time dimension to be included and policy-related data sets. This produces a map of the likely distribution of future afforestation which can be summarized using overlay techniques against a variety of spatial units. Inductive spatial modelling is used to predict habitat suitability for red deer; overlay of the afforestation prediction allows impact of land use change to forestry on red deer habitat to be assessed. Restructuring the forestry prediction around subcatchments of the Rivers Dee and Don illustrates how new analyses can be carried out on output from GIS towards identifying the different impact of future afforestation on the two catchments.

Predicting the location and effect of future afforestation in the Grampian region

A great variety of factors influence the likelihood of afforestation. These include the physical environmental conditions as they influence yield, length of rotation, and windthrow, the operation of constraints such as those summarized within an Indicative Forestry Strategy (Strathclyde Regional Council, 1988), the economics of forestry and alternative land uses, the operation of a range of policies, and individual decisions of land owners and investors. At a broad *regional* level some of these influences can be considered within a simple model. Since the model is simple, its assumptions and limitations should be noted; it does, however, provide a general indication of *where* change may be expected, establishing the limits rather than the detail. Economic and individual decisions are not considered.

A number of considerations impact on likelihood of afforestation.

1. There has been a recent change of attitude toward conversion of agricultural land to other uses (Scottish Development Department Circular 18/1987, *Development Involving Agricultural Land*). Specifically there is no longer the presumption that land of Land Capability for Agriculture (LCA) classes 3.2, 4.1 and 4.2 shall be retained in agricultural production. Land of LCA classes 1, 2 and 3.1 retains the protection of a presumption for agricultural production (Scottish Development Department, 1981).

2. The principal alternative use of agricultural land is considered to be forestry (Scottish Development Department, 1987) the Government declaring an annual planting target of 33 000 ha, in addition to the 3-year total of 36 000 ha within the Farm Woodlands Scheme.
3. Future upland afforestation is to be confined to Scotland (Strathclyde Regional Council, 1988; Shucksmith, 1988).

These three policy considerations are included within a model of likelihood of future afforestation through applying weightings to LCA classes according to the relative ranking of the classes (Table 16.1).

The environmental input to the model is formed by a geographic model. Tree growth is affected by a wide range of environmental conditions, including climate and soils. Within the Grampian region, much of the observed variation in climate, soil and vegetation type can be described by altitude (Walker *et al.*, 1982) and so this variable is used as a surrogate geographic measure describing environmental conditions and tree growth in the area. As for LCA, each altitude class is weighted, in this case according to the relative utility of the class for forestry (Table 16.1). These allocation rules are then applied using GIS to combine the LCA and altitude data sets to produce an output reflecting the likelihood of future afforestation across the Grampian region (Figure 16.3). This map output can be summarized according to districts (Figure 16.4) or other spatial units using overlay and appropriate area boundaries. It can be seen that operating this model in the Grampian region predicts future afforestation to be concentrated in a broad belt across the region, the majority being in only two of the five districts of the region.

Modelling habitat suitability for red deer

Habitat suitability for red deer has been modelled using an inductive learning process whereby data on the distribution of red deer are used within analysis of other geographic data sets to produce input weightings for data set combination.

The distribution of red deer (*Cervus elaphus* L.) in Deer Management Group areas within the Grampian region is censused by the Red Deer Commission. Counts

Table 16.1. Rule base for forestry prediction.

Land capability for agriculture		Altitude	
Class	Rank	Range (m)	Rank
2	9=	0–60	6=
3.1	9=	61–120	6=
3.2	3	121–180	5
4.1	2	181–240	3
4.2	1	241–300	1
5.1	4	301–420	2
5.2	5	421–600	4
5.3	7	601–900	6=
6.1	6	>900	6=
6.2	8		
6.3	9=		
7	9=		

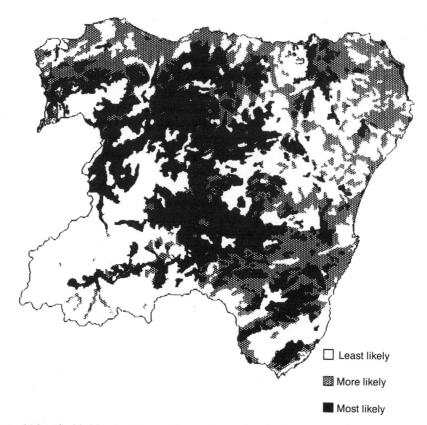

Figure 16.3. The likelihood of future afforestation within the Grampian region.

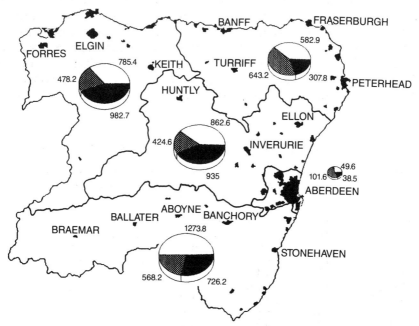

Figure 16.4. The likelihood of future afforestation within the Grampian region—district summaries.

are made during the late winter/early spring when the ground is snow covered. Their number is recorded for each 1 km × 1 km grid square of the National Grid. In the Grampian region, the principal areas where red deer are counted are Glenfiddich, Glen Avon and Upper Deeside. The data give an indication of the winter range of red deer, although a number of management factors influence the distribution of deer, most notably the presence of deer fences which restrict movement to lower altitudes (Red Deer Commission, 1981), and feeding stations such as Mar Lodge and Glen Muick, where densities tend to be artificially high. Locally the distribution of deer is a response to social behaviour (Clutton-Brock *et al.*, 1982), the species composition and nutritional quality of herbage, the need for shelter, and disturbance impact (Mitchell *et al.*, 1977). At a regional scale the major factors controlling distribution are climatic and general habitat features (Mitchell *et al.*, 1977). This analysis concentrates on a regional scale as the most appropriate both for the available environmental data and for comparison with the forestry model.

Three datasets are used in combination with red deer distribution: altitude, land cover and accumulated frost (Table 16.2). Altitude is recorded in nine classes, the minimum contour interval being 60 m with a larger interval (120 m) at higher altitudes. Land cover, which partially reflects land use, is recorded in nine classes, these being obtained from classification of Landsat MSS data (Hubbard and Wright, 1982). Red deer feed on heather and grass but have a preference for *Calluna* vegetation containing grasses rather than for grass-dominated *Agrostis–Festuca* swards (Mitchell *et al.*, 1977). Broad heath, upland scrub and grass-dominated cover types are readily distinguished using LANDSAT MSS data and the overall classification accuracy of the image classification used is high. The third data set, accumulated frost, defined as the annual integrated deficiency of temperature with reference to 0°C (Shellard, 1959), is mapped in five classes at a scale of 1 : 625 000 (Birse and Robertson, 1970). This gives an indication of winter severity.

An index of habitat preference is calculated for each category in the data sets after the method described by Duncan (1983). The percentage of a habitat attribute in the 1 km × 1 km squares occupied by red deer is divided by the percentage of that attribute in the area being considered, in this case Gordon and Kincardine and Deeside districts. The value is transformed by adding one and taking the logarithm (base 10), giving a minium value of 0.0 and providing an index of the 'importance' of each category to the distribution of the species. Values less than 0.3 ($\log_{10}$ 2.0) reflect

Table 16.2. Habitat preference indices for data sets used to predict the distribution of red deer.

Altitude		Land cover		Accumulated frost	
Range (m)	Index	Type	Index	Days-°C	Index
0–60	0.00	Heath	0.49	20–50	0.00
61–120	0.00	Scrub	0.45	51–110	0.00
121–180	0.01	Grass	0.37	111–230	0.40
181–240	0.14	Rock	0.35	231–470	0.69
241–300	0.28	Woodland	0.32	>470	0.14
301–420	0.51	Urban	0.12		
421–600	0.54	Crops	0.02		
601–900	0.58	Sand	0.00		
>900	0.09	Open water	0.00		

a category which is under-represented, suggesting avoidance (although deer cannot strictly be avoiding habitat types in which they occur). Values close to 0.3 reflect occupancy approximately in proportion to availability and is expected if there is neither preference nor avoidance for the category, that is, if use is random. Values in excess of 0.3 show over-representation of the category and indicate preference—in this case, the higher the value, the greater the preference.

Habitat suitability model

Habitat preference indices for each altitude, land cover and accumulated frost class are combined to model habitat suitability for red deer. Each category of the three data sets is weighted according to its habitat preference index, the data sets are overlaid and the habitat suitability class is assessed according to the rules in Figure 16.5. The map output (Figure 16.6) is compared with the distribution of red deer in Moray district to provide an accuracy assessment.

Habitat preference indices for red deer are shown in Table 16.2. The preferred altitudes are 421–600 and 301–420 m (in decreasing order). All other altitude classes have habitat preference indices of less than 0.3. Four cover types have preference indices which are greter than 0.3, namely heath, upland scrub, woodland and rock. Cover types form a complex mosaic within the region and each 1 km × 1 km square may contain more than one type; the relative mix of types is likely to be important. Analysis of the combination of types in 1 km × 1 km squares is not included here. Two of the five classes of accumulated frost recorded in Grampian have indices

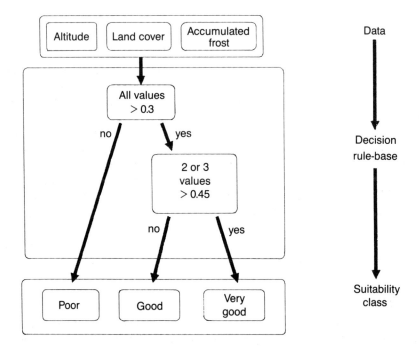

Figure 16.5. Rule base for allocating land with different habitat preference indices to habitat suitability classes for red deer.

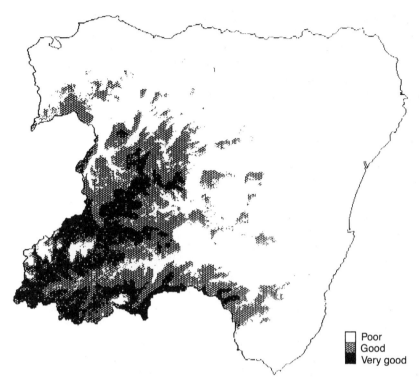

Figure 16.6. Habitat suitability for red deer. The classes are based on applying the rule base shown in Figure 16.5 to the habitat preference indices listed in Table 16.2.

greater than 0.3. The most highly preferred class has between 231 and 470 day-°C below 0°C followed by the milder 111–230 day-°C below 0°C class.

Combining accumulated frost, altitude, and cover type data sets using habitat preference indices to weight the different values of each data set produces a map of habitat suitability for red deer according to these preferences (Figure 16.6). The values in this map are reclassified into three classes in relation to the weightings for the input data sets: two classes represent preferred habitat and one class unsuitable and marginal habitat. Over 75% (6590 km²) of the Grampian region is classed as unsuitable or marginal for red deer, the two classes of preferred habitat covering the remaining 25% (2163 km²). The class representing the most preferred habitat covers some 1320 km². Unsuitable and marginal habitat occurs in the north-east, and towards the coastal areas of the region; preferred habitat is concentrated around the Grampian Mountains in the west of the region, around Lochnagar and Glen Muick, in the Ladder Hills, Hills of Cromdale, Correen Hills, Hill of Fare, on Bennachie and in a belt through Glen Tanar, Forest of Birse and Glen Dye towards the east.

Testing the model
Comparing the habitat suitability map with the distribution of red deer in the Moray district gives encouraging results. Of the 38 1 km × 1 km grid squares in which red deer were recorded the proportion of habitat of the two suitable classes is greater than 50% in every case and in 33 of the 38 squares the proportion is 100%.

Comparing red deer habitat with the forestry model

The red deer habitat model can be compared with the predicted distribution of future afforestation using GIS to overlay the maps to identify areas where land-use change may affect red deer distribution or, alternatively, where red deer may come into conflict with forestry. In this case less than 2.5% of 'good' or 'very good' red deer habitat falls within the areas most likely to experience afforestation (Figure 16.7). A similar approach can be adopted for other species and has particular potential for application in establishing conservation interest on a regional scale and in areas outside nature reserves.

Further analysis

Output from the forestry model may also be *restructured* around any specified spatial unit to provide an input to further analysis. Examples of this assessment of the indirect impact of operation of a policy include restructuring output according to agricultural parish boundaries for analysis against the annual agricultural census returns. This allows some assessment of the social and economic impacts of afforestation compared with agricultural uses. Similarly, restructuring against sub-catchments of main rivers provides input to assessment of hydrological and water quality consequences. Figure 16.8 shows the proportion of the area of downstream subcatchments of the Rivers Dee and Don predicted as most likely to be afforested. The Dee is expected to experience relatively less afforestation than the Don, the

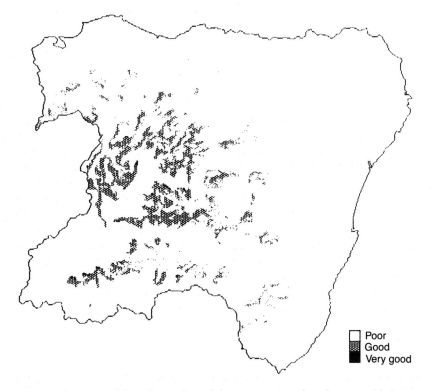

Poor
Good
Very good

Figure 16.7. Habitat suitability classes for red deer in areas predicted as most likely to experience afforestation in the Grampian region (Figure 16.3).

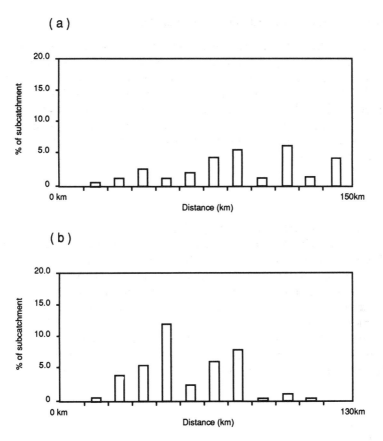

Figure 16.8. Proportion of afforestation classes (Figure 16.3) in downstream subcatchments of the Rivers Dee (a) and Don (b).

amount increasing further downstream. In contrast, the Don catchment is expected to experience extensive afforestation, principally in the upper and middle stretches. This has implications for impact on water and sediment yield and water quality in the river and highlights a greater need for environmentally sensitive planting in the Don catchment in relation to water supplies and the river. The forestry prediction map, restructured around subcatchments, can be further related to soil types, existing land-use and land-management practices and identify areas susceptible to erosion. Models allowing such analyses are currently being developed for application to the Grampian region (MLURI Annual Report 1987). A variety of other analyses are possible which contribute to discussion of indirect consequences of policy. GIS which allow intervisibility analysis and perspective viewing permit landscape impact to be estimated, this often being among the most obvious consequences of forestry. Similarly, social and economic consequenes can be predicted through incorporation of appropriate models with GIS.

Conclusions

The framework for evaluation and assessment of direct and indirect impact of policy using GIS and models allows a wide variety of questions to be asked and reveals a range of possible impacts for consideration. This permits a systematic evaluation of policies and plans and, within the limits of models used, allows a variety of proposals to be compared. Given the role of models in guiding decision-making, the limits of models used need to be made prominent in order that they are not given undue authority and considered a substitute for critical evaluation. Models do, however, provide a form whereby thinking and analysis can be structured and discussion focused. At present there are few models compatible with GIS, and model development is a research priority.

Application of GIS in decision-making through the type of analyses described above is below its potential, both because of a lack of suitable models and because of a lack of appropriate databases and restricted use of GIS technology among those directly concerned with decision-making and policy formulation and evaluation. Appropriate systems and models will rely on collaboration between users and those involved with developing technology.

Notwithstanding these constraints, the potential does exist whereby plans and policies can be formulated on the basis of a wide evaluation of their likely consequences. This may lead to the development of a more integrated approach to land-use planning and, with appropriate developments in systems and modelling, one which is based on a holistic understanding of environment and human activity.

References

Bibby, J. S., Douglas, H. A., Thomasson, A. J. and Robertson, J. S., 1982, *Land Capability Classification for Agriculture*, Aberdeen: Macaulay Institute for Soil Research.

Bibby, J. S., Heslop, R. E. F. and Hartnup, R., 1988, *Land Capability Classification for Forestry in Britain*, Aberdeen: Macaulay Land Use Research Institute.

Birse, E. L. and Robertson, L., 1970, *Assessment of Climatic Conditions in Scotland. 2. Based on Exposure and Accumulated Frost* (map and explanatory booklet), Aberdeen: Macaulay Institute for Soil Research.

Burrough, P. A., 1986, Five reasons why geographical information systems are not being used efficiently for land resources assessment, *Proc. Auto Carto London, Digital Mapping and Spatial Information Systems*, Vol. 2, pp. 139–148, Oxford: Clarendon Press.

Chorley, R., 1988, Some reflections on the handling of geographical information, *International Journal of Geographical Information Systems*, **2**, 3–9.

Cloke, P. A., 1987, Policy and implementation decisions, in Cloke, P. A. (Ed.), *Rural Planning: Policy into Action?*, pp. 19–23, London: Harper & Row.

Clutton-Brock, T. H., Guiness, F. E. and Albon, S. D., 1982, *Red Deer: Behaviour and Ecology of Two Sexes*, Edinburgh: Edinburgh University Press.

Davidson, D. A. (Ed.), 1986, *Land Evaluation*, New York: Van Nostrand Reinhold.

Davidson, D. A., 1990, Perspectives on computer-based techniques in land evaluation, in Bibby, J. S. and Thomasson, A. J. (Eds), *The Evaluation of Land Resources in Scotland*, Proceedings Royal Scottish Geographical Society Annual Symposium, Stirling 1989, Aberdeen: Macaulay Land Use Research Institute.

Department of the Environment, 1987, *Handling Geographic Information*, Report of the Committee of Enquiry chaired by Lord Chorley, London: HMSO.

Duncan, P., 1983, Determinants of the use of habitat by horses in a Mediterranean wetland, *Journal of Animal Ecology*, **52**, 93–109.

Fisher, P. F., 1989, Expert system applications in geography, *Area*, **21**, 279–287.

Hubbard, N. K. and Wright, R., 1982, A semi-automated approach to land cover classification of Scotland from LANDSAT, in *Remote Sensing and the Atmosphere, Proc. Eighth Annual Conference of the Remote Sensing Society*, pp. 212–221, Liverpool: Remote Sensing Society.

Jeffers, J. N. R., 1980, Dynamic modelling of land-use assessment, in Thomas, M. F. and Coppock, J. T. (Eds), *Land Assessment in Scotland*, pp. 119–126, Aberdeen: Aberdeen University Press.

Mitchell, B., Staines, B. and Welch, D., 1977, *Ecology of Red Deer*, Huntingdon: ITE.

Natural Environment Research Council, 1988, *Geographic Information in the Environmental Sciences*, Report of the Working Group on Geographic Information, NERC.

Newson, M. D., 1987, *Land and Water: The 'River Look' on the Face of Geography (Seminar Paper No. 51)*, Newcastle upon Tyne: Department of Geography, University of Newcastle upon Tyne.

Newson, M. D., 1988, *Environmental Use of GIS at the Regional Scale in Britain: Some Methodological and Institutional Constraints.* Northern Regional Research Laboratory Research Paper, Newcastle upon Tyne.

Oliver, M., Webster, R. and Gerrard, J., 1989a, Geostatistics in physical geography. Part I: Theory, *Transactions of the Institute of British Geographers*, **14**, 259–269.

Oliver, M., Webster, R. and Gerrard, J., 1989b, Geostatistics in physical geography. Part I: Applications, *Transactions of the Institute of British Geographers*, **14**, 270–286.

Openshaw, S., Charlton, M., Wymer, C. and Craft, A., 1987, A mark 1 geographical analysis machine for the automated analysis of point data sets, *International Journal of Geographical Information Systems*, **1**, 335–358.

Quarmby, N. A., Cushnie, J. L. and Smith, J., 1988, The use of remote sensing in conjunction with Geographic Information Systems for local planning, *Proc. International Geoscience and Remote Sensing Symposium, Edinburgh 1988*, Vol. 1, pp. 89–92. Remote Sensing Society.

Red Deer Commission, 1981, *Red Deer Management*, Edinburgh: HMSO.

Rhind, D., 1988, A GIS research agenda, *International Journal of Geographical Information Systems*, **2**, 23–28.

Scottish Development Department, 1981, *National Planning Guidelines: Priorities for Development Planning*, Edinburgh: Scottish Development Department.

Scottish Development Department, 1987, *Land Use Summary Sheet—Agriculture* (National Planning Series), Edinburgh: Scottish Development Department.

Shellard, H. C., 1959, Averages of accumulated temperature and standard deviation of monthly mean temperature over Britain, 1921–50, *Professional Notes Meteorological Office No. 125*, London.

Shucksmith, M., 1988, Current rural land-use issues in Scotland: an overview, *Scottish Geographical Magazine*, **104**, 176–180.

Sivertun, A., Reinelt, L. E. and Castensson, R., 1988, A GIS method to aid non-point source critical area analysis, *International Journal of Geographical Information Systems*, **2**, 365–378.

Strathclyde Regional Council, 1988, *Strathclyde Structure Plan Update (1988) Written Statement*, Strathclyde: Department of Physical Planning, Strathclyde Regional Council.

Walker, A. D., Campbell, C. G. B., Heslop, R. E. F., Gauld, J. H., Laing, D., Shipley, B. M. and Wright, G. G., 1982, *Eastern Scotland: Soils and Land Capability for Agriculture*, Aberdeen: Macaulay Institute for Soil Research.

Walker, P. A. and Moore, D. M., 1988, SIMPLE. An inductive modelling and mapping tool for spatially-oriented data. *International Journal of Geographical Information Systems*, **2**, 347–363.

Worrall, L., 1989, Design issues for planning-orientated spatial information systems, *Mapping Awareness*, **3**, 17–20.

17

Effects of beaver and moose on boreal forest landscapes

C. A. Johnston, J. Pastor and R. J. Naiman

Introduction

Geographical information systems (GISs) have been applied in a number of wildlife habitat studies (Tomlin *et al.*, 1980; Reed, 1980; Donovan *et al.*, 1987; Hodgson *et al.*, 1988; Agee *et al.*, 1989), but these studies have focused on how animals respond to their environment rather than how they *affect* it. Animal effects on the landscape are typically diffuse in both time and space, making them difficult to quantify with or without a GIS. Techniques exist for monitoring animal presence and for mapping environmental conditions, but techniques for monitoring the cumulative effects of an animal population at the landscape scale are generally lacking. Remote sensing and aerial photography can provide mappable evidence of animal influences for GIS analysis, but only if those influences are large and distinctive enough to be detected remotely, and can be positively attributed to animal activities (e.g. insect defoliation outbreaks).

While remote sensing can be used to detect the location and extent of some animal disturbances, field work is usually needed to determine the ecological ramifications of those disturbances. For example, field work would be needed to determine changes in plant standing crop, vegetation cover, biomass partitioning, and ecosystem nutrient dynamics caused by insect outbreaks. Once determined, these data could be combined with GIS-generated data on the areal extent of influence to determine cumulative ecological effects.

When animal activities do not provide a discrete patch, field techniques must be used to both detect and quantify their influences. Fenced exclosures are commonly used to determine the effects of animals on ecosystems (Krefting, 1974; Hanley and Taber, 1980; Hatton and Smart, 1984; McNaughton, 1985). The exclosures are typically designed to exclude a specific animal species, and are left in place sufficiently long to determine the effects of animal activities on vegetation or other ecosystem properties. A GIS could be useful in extrapolating data from animal exclosures to the landscape as a whole, provided that the animal influence is uniform across the landscape or its distribution is known (e.g. from radiotelemetry data).

We have used a combination of GIS, air-photo interpretation, and field techniques to determine the influence of beaver ponds on landscape processes in the 298 km^2 Kabetogama Peninsula of Voyageurs National Park, 25 km east of International Falls, Minnesota, USA (Naiman *et al.*, 1988; Broschart *et al.*, 1989; Johnston and Naiman, 1987, 1990b, c). We have also studied the influence of animal herbivory on forest ecosystems on Isle Royale, Michigan, USA (Pastor *et al.*, 1987,

1988), and at beaver ponds near Duluth, Minnesota (Johnston and Naiman, 1990a), and are in the process of combining these findings with GIS-derived data to determine the effects of herbivory at the landscape scale.

The primary objective of the following paper is to examine the effects of two large mammals, beaver (*Castor canadensis*) and moose (*Alces alces*), on boreal forest ecosystems. In doing so, we will illustrate how GIS was an indispensible tool for extrapolating field data to the landscape as a whole, and for deriving landscape ecological relationships which would have been impossible using traditional field methods. In addition, we will discuss potential ways in which GIS could be used in future research on animal/forest interactions.

Alteration of water flow

Beaver dams alter the aquatic landscape by increasing water depth and surface area while decreasing the flow rate of streams (Naiman *et al.*, 1988). Impoundments provide habitat for beaver by affording protection from predators (primarily wolf, *Canis lupus*), increasing accessibility to riparian food supplies, and facilitating trans-

Table 17.1. GIS procedures used to characterize beaver impoundments and measure transition rates. (Reproduced with permission from Johnston and Naiman (1990b).)

Step	Description	ERDAS programs used
A.	Measure beaver impoundment area by vegetation and hydrologic type	
	1. Digitize peninsula outline, lakes, and impoundment types	DIGPOL
	2. Create background file of the peninsula and its permanent lakes	MAKEFIL, GRDPOL, CUTTER
	3. Create beaver impoundment file, detailed classification	MAKEFIL, GRDPOL, CUTTER
	4. Overlay beaver impoundment file on peninsula background	OVERLAY
	5. Recode file to generalized landscape types	RECODE
	6. Measure total area of each landscape type	BSTATS
	7. Export area data to ASCII file	EXPTRL
B.	Compute transition rates and error check maps	
	1. Compute transition rates for each possible change, and recode output	MATRIX
	2. Execute error-checking routine to locate illogical transitions	CLUMP, SIEVE
	3. Make corrections to maps as needed, repeat steps (A1) through (B1), export data	EXPTRL
C.	Define and measure individual impoundments and impoundment clusters	
	1. Digitize primary pond watersheds	DIGPOL
	2. Define individual ponds by intersecting watershed and impoundment files, export data	RECODE, MATRIX, EXPTRL
	3. Define and number impoundment clusters, export data	RECODE, CLUMP, SIEVE, EXPTRL
D.	Measure length of streams impounded by beaver	
	1. Digitize streams by stream order	DIGPOL
	2. Create raster file of streams	MAKEFIL, GRDPOL
	3. Overlay impoundment and stream files, export data	RECODE, OVERLAY, EXPTRL
	4. Multiply number of pixels in impounded stream by conversion coefficient	See text

port of woody materials used for construction and food (Johnston and Naiman, 1987).

The relatively large size (>0.5 ha) of most beaver ponds, and their sharp contrast with the surrounding forest, makes it possible to accurately quantify the effect of beaver on landscape hydrology using aerial photography. We mapped and classified by cover type all areas in which vegetation had been altered by beaver impoundments, using aerial photos taken in 1940, 1948, 1961, 1972, 1981 and 1986. The 1 : 24 000 maps were digitized and analysed using an ERDAS raster-based GIS, which was used to analyse the aerial extent, distribution and characteristics of the beaver ponds over time (Table 17.1). The following GIS data layers were manually digitized from the topographic and beaver pond maps using a Calcomp 91480 digitizing table: peninsula outline, permanent lakes within the peninsula, beaver pond outlines, wetland types within beaver ponds for each date of photography, drainage basins of primary ponds, areas burned by 1936 fire, and areas logged between 1950 and 1970. Since ERDAS is a raster-type GIS, the digitized polygons were gridded into a matrix of 7 m × 7 m picture elements (i.e. pixels), a pixel size determined experimentally to be appropriate for the data resolution. GIS-derived data from each of the six air-photo dates were exported to a spreadsheet program (LOTUS 123), which was used to compute summary statistics for each date and trends for four time periods between aerial photo dates: 1940–1948, 1948–1961, 1961–1972, 1972–1981, which approximate decadal intervals (Johnston and Naiman, 1990b).

Our GIS data showed that there were few beaver ponds on the Kabetogama Peninsula in 1940, a time when trapping and habitat destruction had decimated the beaver population (Figure 17.1(a)). By 1986, however, there were 2.5 dams per kilometre of stream, and dam frequency in the landscape was 2.8/km^2 (Naiman *et al.*, 1988). The proportion of stream length impounded by beaver in the Shoepack Lake drainage basin, a 38 km^2 portion of the Kabetogama Peninsula, was 53% for first-order streams, 55% for second-order streams, and 87% for fourth-order streams (Johnston and Naiman, 1990b). The third-order 'streams' were two lobes of Shoepack Lake, and were, therefore, unimpoundable.

Summary statistics generated using the GIS showed that a substantial portion of the landscape was hydrologically altered by beaver. Beaver ponds which had been created by 1986 covered a cumulative area of 37.7 km^2, 13% of total peninsula area (Naiman *et al.*, 1988; Johnston and Naiman, 1990c). These beaver impoundments increased the area of aquatic habitat, which in the absence of beaver was primarily lake area, by 438%. Therefore, beaver converted the peninsula from a landscape dominated by forest to a spatial mosaic of aquatic and terrestrial habitats.

Patch creation

Physical and biological disturbances in the landscape produce patches, discrete communities embedded in an area of dissimilar community structure or composition (Pickett and White, 1985). The alteration of edaphic conditions by beaver ponds creates spot disturbance patches in a matrix of upland forest (Remillard *et al.*, 1987; Johnston and Naiman, 1990c). Where before there was only a narrow stream corridor among the forest matrix, a beaver dam creates a patch of water with very different properties than the stream and forest it replaced (Figure 17.2).

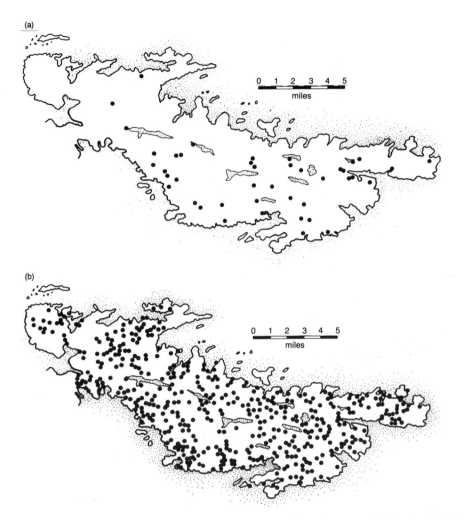

*Figure 17.1. Open water areas created by beaver dams on the Kabetogama Peninsula: (a) 1940,
(b) 1986.*

As beaver harvest trees and shrubs in the riparian zone surrounding their pond,
a second patch develops concentric to the first (Figure 17.2). Beaver are central
place foragers (Orians and Pearson, 1979), so their feeding activities are confined to
the riparian zone within about 100 m of their pond (Jenkins, 1980; McGinley and
Whitham, 1985; Basey *et al.*, 1988). Riparian zones dominated by the deciduous
species preferred by beaver, such as trembling aspen (*Populus tremuloides*), may be
virtually clear-cut (Johnston and Naiman, 1990a). The stand becomes more open,
and shrubs and root suckers become the dominant growth form (Figure 17.2).

The boreal forest presents a paradox to moose because they need a spatially
diverse landscape that provides cover as well as food (Telfer, 1984). While the preferred foods are early successional hardwoods (e.g. trembling aspen, balsam poplar
(*Populus balsamifera*), and paper birch (*Betula papyrifera*)), which can only be found
within browse height in recently disturbed patches, the preferred cover is mature
late successional conifers, such as white spruce (*Picea glauca*), which is almost never
browsed. Our studies (Pastor *et al.*, 1988; McInnes *et al.*, 1992) have shown that

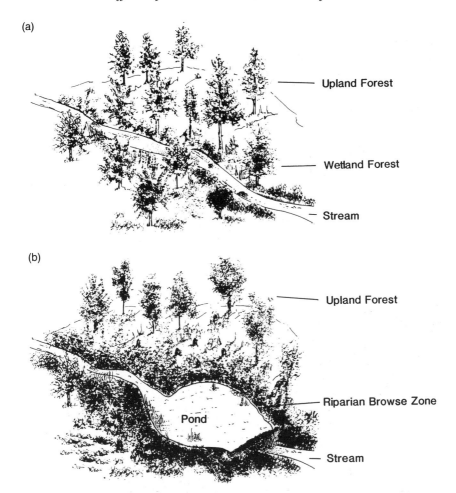

Figure 17.2. Landscape patches created by beaver impoundments and foraging. (a) Before and (b) after beaver colony establishment.

heavy moose browsing on young early successional hardwoods can prevent the development of mid successional closed canopy forests, substituting instead a 'spruce–moose savanna'. Typically, spruce assumes dominance in a two-step fashion, whereby established seed trees serve as foci for 'infection' of the surrounding area by their progeny. Consequently, the spruce–moose savanna is a patchy environment of spruce groves separated by patches of heavily browsed forage. Others have shown that the intensity of moose browsing decreases with increasing distance from cover (Hamilton *et al.*, 1980).

On a larger scale, Geist (1974) hypothesized that moose find refuge in habitats of small but continuous disturbance, such as river floodplains, but expand into transient habitats when large disturbances such as fire create an abundance of food. In eastern boreal forests and in the Rocky Mountains, beaver ponds and their associated foraging areas are important small but continuous disturbances that are often frequented by moose because of their abundant aspen, birch and hazel browse (Rudersdorf, 1952; Jonas, 1955, 1959; Wolfe, 1974). However, heavy browsing by moose on small aspen may prevent them from attaining the larger diameters preferred by beaver. Thus, patch creation by beaver benefits moose, but the subsequent influence of moose on the dynamics of those patches may be detrimental to beaver (Figure 17.3).

In the absence of repeated disturbance, patches usually revert to conditions found in the surrounding forest matrix. Forman and Godron (1986) have called this 'patch extinction'. On the Kabetogama Peninsula, however, all beaver impoundments that had been created on the Kabetogama Peninsula during the 46-year study peiod were still clearly distinguishable on the 1986 aerial photos. Although some areas were briefly abandoned, they were either reflooded by new beaver occupants or had been altered to such an extent as to retard secondary succession. Therefore, there was no patch extinction over the time period studied, and impoundment numbers and area were cumulative (Johnston and Naiman, 1990b, c).

The extreme longevity of beaver-created landscape patches is not common to all disturbance patches on the Kabetogama Peninsula. A severe fire burned much of the peninsula in 1936, and pulp logging was done in selected areas between 1950 and 1970 (Rakestraw et al., 1979). Although the fire and pulp logging respectively affected 56% and 12% of peninsula area, the patches which resulted had all reverted to forest cover by 1986. Woody revegetation of former beaver ponds, in contrast, occurred in less than 5% of the area impounded (Johnston and Naiman, 1990b).

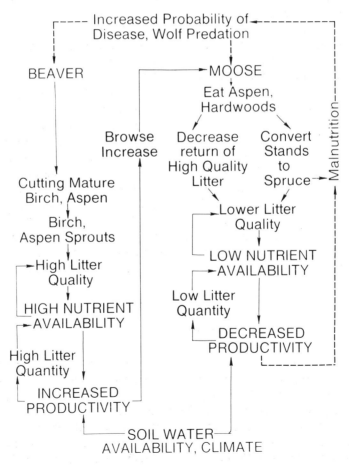

Figure 17.3. Interactions of moose, beaver, and wolves with vegetation and soil properties.

The longevity of beaver pond patches relative to other disturbance patches is probably due to the severity of the disturbance (i.e. conversion of terrestrial soils to aquatic soils, death and replacement of both overstorey and understorey vegetation) and the rapid return interval between disturbance events (10–30 years) (Remillard *et al.*, 1987).

Patch creation by physical disturbances is often stochastic, occurring at unpredictable time intervals. Patch creation by animal disturbance is less stochastic because it is related to numbers of animals present, which change over time in response to fairly predictable patterns of birth, death, and migration (Botkin *et al.*, 1981) and the criteria by which they select food and habitat (Senft *et al.*, 1987). We have used GIS techniques to relate rates of pond establishment and vegetation change on the Kabetogama Peninsula to beaver population data (Broschart *et al.*, 1989; Johnston and Naiman, 1990c). Beaver numbers were very low in the 1940s due to overtrapping, but increased to about 1 colony/km^2 by 1986 (Smith and Peterson, 1988). Although the beaver population was variable from year to year, there was an overall increase of about 9 colonies/yr (Broschart *et al.*, 1989). After 1961, the beaver population trend paralleled the increase in number of impoundments (9.8 new impoundments/year), indicating that the establishment of new ponds was driven by the increase in beaver numbers (Johnston and Naiman, 1990c). Prior to 1961, however, the increase in number of beaver pond sites (25.2 new impoundments/year) greatly exceeded the increase in beaver population. These data demonstrate that beaver can quickly create new patches through their dam-building activities, rapidly spreading into suitable habitats as their population increases.

Where animal numbers are increasing, more favourable home ranges are often occupied first (Hunter, 1964; Senft *et al.*, 1987). Low-status animals and juveniles may be forced into low-quality sites, resulting in densities higher than expected from forage availability (Van Horne, 1983). We believe this explains the higher than expected pond creation rates on the Kabetogama Peninsula prior to 1961, since juvenile beavers are forced out of their parent colony at about 2 years of age in search of new habitat (Jenkins and Busher, 1979). After 1961, new beaver pond establishment was constrained by a geomorphic threshold (Swanson *et al.*, 1988) between the availability of easily floodable versus marginally floodable sites for pond establishment (Johnston and Naiman, 1990c). Because beaver impoundments are restricted to portions of the landscape in which a low dam can retard the flow of water sufficiently to create a pond, there is a limit to the area which can potentially be affected by beaver. Once this limit is attained, the area affected tends to remain constant due to the extreme longevity of beaver impoundment patches.

Ecosystem effects of impoundments

Because they are created and maintained by living organisms, beaver ponds are themselves dynamic, changing as they are colonized, flooded, and abandoned by beavers. The ponding of water kills vegetation intolerant of flooding, replacing terrestrial with aquatic vegetation. After the pond is abandoned, the dam deteriorates, the water drains, and the exposed sediments are revegetated by water-tolerant grasses and sedges. Eventually, the site is recolonized by beaver, completing the cycle of beaver pond succession (Naiman *et al.*, 1988). At any point in time, beaver-occupied landscapes are therefore a mosaic of different vegetation types, ranging

from submersed aquatic vegetation in deep water areas to lowland forest in seasonally flooded areas (Figure 17.4). Using the US Fish and Wildlife Service wetland
classification system (Cowardin *et al.*, 1979), we identified 32 different cover types in
beaver ponds of the Kabetogama Peninsula. Open water constituted 27% of the
area impounded in 1986, with herbaceous and woody areas constituting 44% and
29% of impounded area, respectively (Johnston and Naiman, 1990b).

Since beaver ponds are not static, however, it was necessary to quantify vegetation and hydrologic changes in beaver ponds over time. This was done with the GIS
by comparing beaver impoundment data layers from the onset versus the end of
each decadal period (Johnston and Naiman, 1990b). The major change in the 1940s
was the creation of new beaver impoundments (Figure 17.5), which rapidly increases
the total area impounded. Nearly 80% of the cumulative area affected by beaver in
the 1940s involved the conversion of upland to ponds, emergent marshes, and
flooded woods. During the 1970s, new impoundments constituted only 9% of the
changes, while 25% of the impoundments changed from one vegetation type at the
onset of the decade to another by the end (Figure 17.5). The majority of the
impoundments, however, remained in the same vegetation category over the 10-year
period (Naiman *et al.*, 1988; Johnston and Naiman, 1990b).

Accompanying these vegetation and hydrological trends were changes in soil
and nutrient cycling properties. Along the moisture gradient created by beaver
impoundments, soil redox potential (Eh) increased from an average of -180 mV in
pond sediments to oxidizing values (>500 mV) in unimpounded forest soils
(Naiman *et al.*, 1988). Because nitrogen dynamics are closely tied to Eh, areas with
different flooding regimes also had different standing stocks and forms of nitrogen.
Flooding of soil caused the initial rapid loss of nitrate (NO_3–N) through denitrifica-
tion, the conversion of nitrate to gaseous forms of nitrogen. The anaerobic conditions also blocked nitrification (the conversion of NH_4–N to NO_3–N), which
resulted in the accumulation of ammonium (NH_4–N). This accumulation resulted in
high concentrations of plant-available nitrogen (KCl extractable nitrogen plus dis-
solved nitrogen in soil solution) in ponded soils, where available nitrogen was
29.8 kg ha^{-1} as opposed to only 6.8 kg ha^{-1} in forest soils (Table 17.2).

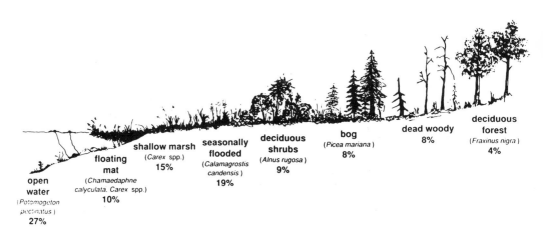

*Figure 17.4. Proportion of beaver impoundments in major vegetation types, 1986. Representa-
tive species are shown in parentheses.*
(Reproduced with permission from Johnston and Naiman (1990b).)

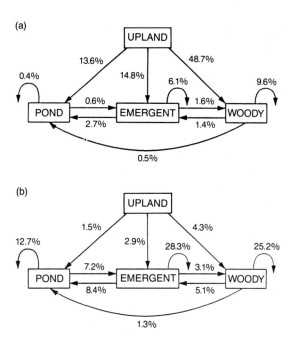

Figure 17.5. Impoundment conversions, by vegetation type. Expressed as percentage of total area impounded at the end of the decade. New impoundments are indicated by arrows emanating from the upland box. Conversion rates are compared for the 1940s (a) and 1970s (b) for the four principal community types.
(Reproduced with permission from Naiman et al. (1988).)

Table 17.2. Comparison of vegetative-hydrological cover types, available forms of soil/sediment nitrogen and total nitrogen for 1940 and 1986 on the Kabetogama Peninsula, Minnesota. (Reproduced with permission from Naiman et al. (1988).)

Cover type	Area (ha)	Concentration (kg ha^{-1})		Absolute amounts (kg)	
		Total nitrogen	Available nitrogen	Total nitrogen	Available nitrogen
1940					
Forest	3508	3050	6.8	5.34×10^6	11.92×10^3
Moist	215	3216	7.7	0.35×10^6	0.83×10^3
Wet	28	2912	16.2	0.04×10^6	0.23×10^3
Pond	16	3543	29.8	0.03×10^6	0.24×10^3
Total	3767			5.76×10^6	13.22×10^3
1986					
Forest*	0	3050	6.8	0	0
Moist	1367	3216	7.7	2.20×10^6	5.26×10^3
Wet	1029	2912	16.2	1.50×10^6	8.33×10^3
Pond	1371	3543	29.8	2.43×10^6	20.41×10^3
Total	3767			6.12×10^6	34.01×10^3

* For 1986 we assume the area affected by beaver is a maximum. Thus, no additional forest area will be affected.

Extrapolating these nutrient concentration data to the peninsula as a whole using the GIS demonstrated that these sediment-level effects have landscape-level implications (Naiman *et al.*, 1988). Although the conversion of forests to aquatic ecosystems by beaver has not changed total nitrogen standing stocks, it has more than doubled the standing stocks of available nitrogen in areas impounded by beaver between 1940 and 1986 (Table 17.2).

Effects of herbivory on forest composition

Large animal herbivory can significantly influence ecosystem structure and dynamics. Hatton and Smart (1984), Tiedemann and Berndt (1972), and Cargill and Jefferies (1984) found larger pool sizes of soil nutrients or greater cation exchange capacity inside elephant (*Loxodonta africana*), elk (*Cervus canadensis*), and snow goose (*Chen hyperborea*) exclosures, respectively, compared with outside. While herbivory of herbaceous plant species can decrease vegetation cover and species richness (Huntly, 1987), it can increase aboveground primary productivity, preventing tissue senescence and maintaining plant nutrient status (McNaughton, 1985). Herbivory is especially significant in boreal forests because limited light, temperature, and nutrient resources restrict the capacity of woody plants to replace tissues eaten by browsing animals (Bryant and Chapin, 1986).

Although moose and beaver both prefer early successional hardwoods such as trembling aspen and paper birch (Peterson, 1955; Northcott, 1971; Krefting, 1974; Risenhoover and Maas, 1987; Johnston and Naiman, 1990a), there are many differences between moose and beaver browse patterns. These differences have important consequences for ecosystem functioning. While moose herbivory is restricted to the forest understorey, beavers can fell mature trees, effectively allowing them to 'browse' the forest canopy. Moose herbivory removes woody tissues which can be replaced by the plant within several growing seasons through new twig or seedling growth, while beaver herbivory removes the entire above-ground biomass of the trees which they fell. Moose consume all the biomass that they browse, while beaver only consume a fraction, leaving most of the bole and branches on the ground (Aldous, 1938). Beavers are central place foragers restricted to a small home range (Jenkins, 1980), while the browse effects of an individual moose may be scattered over hundreds of kilometres (Phillips *et al.*, 1973). Therefore, beaver and moose herbivory may have very different effects on forest standing crop, biomass partitioning, and patchiness (Johnston and Naiman, 1990a).

There are several reasons why beaver have a much greater potential than other large animals to alter ecosystems through herbivory. First, as the only animals in North America except humans which can fell mature trees, their ability to decrease forest biomass is much greater than that of other herbivores. Second, the impact of beaver foraging is concentrated within a small area (Johnston and Naiman, 1987). Finally, the amount of biomass harvested by beaver far exceeds the amount actually ingested, so that nutrients remain locked in slowly decomposing woody tissues rather than returning to the soil in the form of excrement (Johnston and Naiman, 1990a).

We quantified the effects of beaver herbivory on biomass at our beaver pond study sites near Duluth, Minnesota (Johnston and Naiman, 1990a). Each beaver felled an average of about 1.4 Mg per year of woody biomass within the 1 ha browse zone. A moose, in comparison, browses only 0.0003–0.091 Mg ha^{-1} year^{-1} of woody biomass within its foraging area (Crête and Bédard, 1975; Bédard *et al.*, 1978; Crête and Jordan, 1982; Risenhoover, 1987; McInnes, 1989). The beaver colony at Ash Pond harvested nearly twice as much biomass per unit area as a herd of Serengeti ungulates (4.38 Mg ha^{-1} year^{-1}) (McNaughton, 1985). Therefore, despite its small size, an individual beaver appears to have a much larger potential effect on stand biomass within its foraging range than any other browser (Johnston and Naiman, 1990a).

Long-term foraging by beaver significantly altered tree density and basal area at the Duluth beaver ponds. At Ash Pond, where beaver had been foraging uninterrupted for six years, the density and basal area of stems ⩾5 cm were reduced by 43%. The decrease in stand density (7%) and basal area (20%) was much less at Arnold Pond because of the shorter duration of beaver herbivory (Johnston and Naiman, 1990a).

The reduction in overstorey density caused by the removal of large trees by beaver at the two study sites has a number of potential ramifications to biomass partitioning. By increasing light penetration, the productivity of existing subcanopy trees and shrubs should increase. By promoting suckering and seed germination, reproduction rates should increase. Both of these effects should increase the proportion of understorey biomass. Increased understorey biomass could be offset, however, by increased herbivory by other boreal forest browsers (e.g. moose and hare) if the lowering of average canopy height increases the amount of accessible woody biomass.

Because beaver selected certain species over others, the residual overstories at both ponds had a different species composition than their pre-browse overstories (Johnson and Naiman, 1990a). At Ash Pond, the loss of trembling aspen was offset primarily by an increase in the importance value of black ash and tag alder (Table 17.3). This increase does not mean that these species became more abundant, only that they constituted a larger proportion of the residual canopy than the pre-browse canopy. The loss of trembling aspen at Arnold Pond resulted in small increases in

Table 17.3. Changes in species importance values caused by beaver browsing stems ⩾5 cm in diameter.

Species	Arnold Pond	Ash Pond
Abies balsamea	+0.7	—
Acer rubrum	—	+0.1
Acer saccharum	+0.1	+1.5
Acer spicatum	+0.1	0
Alnus rugosa	+1.4	+7.2
Amelanchier spp.	0	+0.6
Betula papyrifera	+1.4	+0.1
Cornus stolonifera	—	—
Corylus cornuta	—	—
Fraxinus nigra	0	+8.9
Picea glauca	+1.4	—
Populus balsamifera	+1.3	−0.4
Populus tremuloides	−7.4	−17.7
Prunus spp.	+0.1	−2.6
Salix spp.	+0.9	+1.0
Ulmus americana	—	+1.4

the importance of tag alder, paper birch, white spruce, and balsam poplar (Table 17.3).

We hypothesize that long-term beaver herbivory (i.e. several decades) will cause substantial changes in forest composition due to the preferential selection of some browse species over others by beaver (Figure 17.6). Based on the trends we have observed at Ash and Arnold ponds and at Voyageurs National Park, we expect short-term browsing by beaver to cause a shift from aspen dominated stands to stands dominated by less preferred deciduous species such as balsam poplar and birch, similar to the predictions of Barnes and Dibble (1988) for beaver-browsed lowland hardwood forests. Longer term browsing of deciduous stands would cause them to become replaced by shrub zones of unpalatable and/or highly competitive species, such as tag alder or beaked hazel. Eventually, preferential herbivory would increase the competitive advantage of slowly growing, unpalatable conifers (Bryant and Chapin, 1986; Pastor *et al.*, 1988). Thus, preferential central place foraging by beaver scattered throughout the landscape could establish a patchwork of shrubs and conifers in an otherwise deciduous forest matrix, increasing forest diversity.

While the effect of beaver herbivory is large within the area surrounding an individual pond, it is less so at the landscape scale because the distribution of beaver is limited by their habitat requirements to areas within about 100 m of a water body (Jenkins, 1980; McGinley and Whitham, 1985). Using the GIS to generate a 100 m foraging zone around each beaver pond, we have estimated that 12–15% of the Kabetogama Peninsula is affected by beaver herbivory (Naiman *et al.*, 1988). A browse rate of 1.4 Mg ha^{-1} year^{-1} within this area would result in a landscape level browse rate of 0.17–0.21 Mg ha^{-1} year^{-1}. This illustrates the value of using a GIS to extrapolate field data to the landscape scale: without the GIS data on

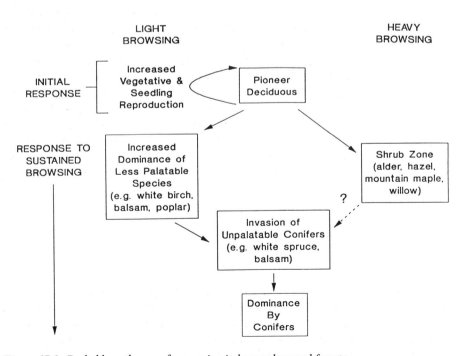

Figure 17.6. Probable pathways of succession in beaver-browsed forests.

approximate foraging area, we might have grossly overestimated the impact of beaver herbivory on the landscape.

Effects of herbivory on nutrient cycling

The alteration of forest species composition by beaver or moose herbivory could have nutrient cycling consequences because of the relationship between litter decomposition rates and nutrient mineralization (Melillo *et al.*, 1982; Flanagan and VanCleve, 1983). Herbivores, vegetation, and soil microbes that decompose organic matter are three interacting parts of feedback loops with both positive and negative components (Figure 17.3). Browsing increases the amount of spruce and fir, whose litter is resistant to decomposition due to recalcitrant carbon compounds and low nutrient content and would thereby decrease soil nutrient availability (Pastor *et al.*, 1988). The rate of nutrient release from litter affects herbivores by controlling the supply of browse and the rate at which plants recover from browsing (Bergerud and Manual, 1968; Peek *et al.*, 1976; Botkin *et al.*, 1981; Pastor *et al.*, 1988). Because of this relationship between feeding preference and litter quality, herbivores should affect soil nitrogen availability and microbial communities to the extent that they alter plant community composition.

On Isle Royale, Pastor *et al.* (1987) showed that percentage carbon and nitrogen in the soil, and nitrogen mineralization were highest in an area where beaver herbivory maintained the presence of mountain ash (*Sorbus americana*), a deciduous species with easily decomposable litter. Beaver foraging significantly elevated soil carbon and nitrogen pools over that found in an adjacent spruce–fir forest without moose or beaver. The increase in relative importance of tag alder (*Alnus rugosa*) observed at the Duluth beaver ponds (Table 17.3), may even have increased total nitrogen inputs to the ecosystem. Alder is a nitrogen fixing species which can add $85–168$ kg N ha^{-1} year^{-1} to the soil (Daly, 1966; Voigt and Steucek, 1969).

Heavy moose browsing on Isle Royale caused white spruce to dominate, and soil %C, %N, and nitrogen mineralization were lowest in these areas (Pastor *et al.*, 1987). Moose browsing around beaver ponds caused nitrogen availability to decline significantly in comparison to the above example for beaver browsing in the absence of moose (Pastor *et al.*, 1987). Recovery from these effects was very slow: although soil carbon and nitrogen content was significantly higher in moose exclosures than outside of them, nitrogen mineralization rates were significantly different in two out of four exclosures even after four decades of protection from moose herbivory (Pastor *et al.*, 1988).

Conclusions

Beaver and moose are capable of altering the landscape in a variety of ways. They share the ability to affect forest vegetation through herbivory, and prefer many of the same browse species. However, the spatial distribution of their herbivory differs substantially, and must be taken into account when determining the landscape-level implications of their activities. A GIS can provide this spatial analysis capability.

In addition to their affects as herbivores, beaver alter the flow of water in the landscape by constructing dams. The ponding of water changes numerous aspects of the forest ecosystem: vegetation, hydrology, redox potential, microbial processes, and soil nutrient concentrations. We used GIS, either alone or in combination with field- and lab-derived data, to study these alterations: (1) length of streams impounded, (2) the location, areal extent, hydrology, and vegetation type of impoundments, (3) changes in hydrology and vegetation over time as a result of new dam construction and vegetation succession, (4) patch dynamics of pond creation, and (5) changes in nutrient standing stocks in the landscape (Naiman *et al.*, 1988; Broschart *et al.*, 1989; Johnston and Naiman, 1990b, c). Future work will include the use of GIS to analyse the effect of beaver herbivory on forest cover, the relationship between topography and beaver pond creation, and the effect of beaver on carbon storage in landscapes.

While the object of most GIS analysis for natural resource management is to produce graphical output, either to the screen or as a hardcopy map, our use of GIS has focused on its ability to numerically describe spatially distributed features (Johnston, 1989). The capability of GIS to reduce the data in a map to a few summary statistics is one of its greatest attractions to ecological researchers, because those statistics can be used with other analytical techniques to derive mathematical relationships between ecological variables (Berry, 1989).

The increased availability of reasonably priced microcomputer-based GIS has occurred at an opportune time in the development of landscape ecology, with its emphasis on spatial and temporal relationships over large areas and long time periods. While GIS is not a panacea for all landscape ecological problems, we have found it to be an indispensible tool for quantifying animal influences at the land-scale level.

Acknowledgements

Research support from the National Science Foundation (BSR-8516284, BSR-8614960, BSR-8817665, BSR-8906843 and DEB-9119614). Equipment support for the Natural Resources Research Institute GIS Laboratory from the National Science Foundation (DIR-8805437). This is contribution number 76 of the Center for Water and the Environment and 13 of the NRRI GIS laboratory.

References

Agee, J. K., Stitt, S. C. F., Nyquist, M. and Root, R., 1989, A geographic analysis of historical grizzly bear sightings in the North Cascades, *Photogrammetric Engineering and Remote Sensing*, **55**, 1637–1642.

Aldous, S. A., 1938, Beaver food utilization studies, *Journal of Wildlife Management*, **2**, 215–222.

Barnes, W. J. and Dibble, E., 1988, The effects of beaver in riverbank forest succession, *Canadian Journal of Botany*, **66**, 40–44.

Basey, J. M., Jenkins, S. H. and Busher, P. E., 1988, Optimal central-place foraging by beavers: tree-size selection in relation to defensive chemicals of quaking aspen, *Oecologia*, **76**, 278–282.

Bédard, J., Crête, M. and Audy, E., 1978, Short term influence of moose upon woody plants of an early seral wintering site in Gaspé Peninsula, Quebec, *Canadian Journal of Forest Research*, **8**, 407–415.

Bergerud, A. T. and Manuel, F., 1968, Moose damage to balsam fir-white birch forests in Newfoundland, *Journal of Wildlife Management*, **32**, 729–746.

Botkin, D. B., Melillo, J. M. and Wu, L. S.-Y., 1981, How ecosystem processes are linked to large mammal population dynamics, in Fowler, C. F. and Smith, T. D. (Eds), *Dynamics of Large Mammal Populations*, pp. 373–387, New York: Wiley.

Berry, B. K., 1989, Beyond mapping: maps as data—'map-ematics', *GIS World*, Mar., 8, **34**, 36.

Broschart, M. R., Johnston, C. A. and Naiman, R. J., 1989, Predicting beaver colony density in boreal landscapes, *Journal of Wildlife Management*, **53**, 929–934.

Bryant, J. P. and Chapin, F. S. III, 1986, Browsing-woody plant interactions during boreal forest plant succession, in Van Cleve, K., Chapin, F. S. III, Flanagan, P. W., Viereck, L. A. and Dyrness, C. T. (Eds), *Forest Ecosystems in the Alaskan Taiga: A Synthesis of Structure and Function*, pp. 213–225, New York: Springer-Verlag.

Cargill, S. M. and Jefferies, R. L., 1984, The effects of grazing by lesser snow geese on the vegetation of a subarctic salt marsh, *Journal of Applied Ecology*, **21**, 669–686.

Cowardin, L. M., Carter, V., Golet, F. C. and LaRoe, E. T., 1979, *Classification of Wetlands and Freshwater Habitats of the United States. (US Fish and Wildlife Services, Office of Biological Services (FWS/OBS-79/31))*, Washington, DC: US Government Printing Office.

Crête, M. and Bédard, J. M., 1975, Daily browse consumption by moose in the Gaspé Peninsula, Quebec, *Journal of Wildlife Management*, **39**, 368–373.

Crête, M. and Jordan, P. A., 1982, Population consequences of winter forage resources for moose, *Alces alces*, in Southwestern Quebec, *Canadian Field-Naturalist*, **96**, 467–475.

Daly, G. T., 1966, Nitrogen fixation by nodulated *Alnus rugosa*, *Canadian Journal of Botany*, **44**, 1607–1621.

Donovan, M. L., Rabe, D. L. and Olson, Jr, C. E., 1987, Use of geographic information systems to develop habitat suitability models, *Wildlife Society Bulletin*, **15**, 574–579.

Flanagan, P. W. and Van Cleve, K., 1983, Nutrient cycling in relation to decomposition and organic matter quality in taiga ecosystems, *Canadian Journal of Forestry Research*, **13**, 795–817.

Forman, R. T. T. and Godron, M., 1986, *Landscape Ecology*, New York: Wiley.

Geist, V., 1974, On the evolution of reproductive potential in moose, *Naturaliste Canadien*, **101**, 527–537.

Hamilton, G. D., Drysdale, P. D. and Euler, D. L., 1980, Moose winter browsing patterns on clear-cuttings in northern Ontario, *Canadian Journal of Zoology*, **58**, 1412–1416.

Hanley, T. A. and Taber, R. D., 1980, Selective plant species inhibition by elk and deer in three conifer communities in western Washington, *Forest Science*, **26**, 97–107.

Hatton, J. C. and Smart, N. O. E., 1984, The effect of long-term exclusion of large hervibores on soil nutrient status in Murchison Falls National Park, Uganda, *African Journal of Ecology*, **22**, 23–30.

Hodgson, M. E., Jensen, J. R., Mackey Jr, H. E. and Coulter, M. C., 1988, Monitoring wood stork foraging habitat using remote sensing and geographic information systems, *Photogrammetric Engineering and Remote Sensing*, **54**, 1601–1607.

Hunter, R. F., 1964, Home range behavior in hill sheep, in Crisp, D. J. (Ed.), *Grazing in Terrestrial and Marine Environments*, pp. 155–171, Oxford: Blackwell Scientific.

Huntly, N. J., 1987, Influence of refuging consumers (Pikas: *Ochotona princeps*) on subalpine meadow vegetation, *Ecology*, **68**, 274–283.

Jenkins, S. H., 1980, A size-distance relation in food selection by beavers, *Ecology*, **61**, 740–746.

Jenkins, S. H. and Busher, P. E., 1979, *Castor canadensis*, *Mammalian Species*, **120**, 1–8.

Johnston, C. A., 1989, Ecological research applications of geographic information systems, in *GIS/LIS '89 Proceedings*, pp. 569–577, Bethesda, MD: American Society for Photogrammetry and Remote Sensing.

Johnston, C. A. and Naiman, R. J., 1987, Boundary dynamics at the aquatic-terrestrial interface: the influence of beaver and geomorphology, *Landscape Ecology*, **1**, 47–57.

Johnston, C. A. and Naiman, R. J., 1990a, Browse selection by beaver: effects on riparian forest composition, *Canadian Journal of Forest Research*, **20**, 1036–1043.

Johnston, C. A. and Naiman, R. J., 1990b, The use of a geographic information system to analyze long-term landscape alteration by beaver, *Landscape Ecology*, **4**, 5–19.

Johnston, C. A. and Naiman, R. J., 1990c, Aquatic patch creation in relation to beaver population trends, *Ecology*, **71**, 1617–1621.

Jonas, R. J., 1955, A population and ecology study of the beaver (*Castor canadensis*) of Yellowstone National Park, M.Sc. Thesis, University of Idaho.

Jonas, R. J., 1959, Beaver, *Naturalist*, **10**, 20–61.

Krefting, L. W., 1974, The ecology of the Isle Royale moose, with special reference to the habitat, *University of Minnesota Agricultural Experiment Station Technical Bulletin 297*, St. Paul, MN.

McGinley, M. A. and Whitham, T. G., 1985, Central place foraging by beavers (*Castor canadensis*): test of foraging predictions and the impact of selective feeding on the growth form of cottonwoods (*Populus fremontii*), *Oecologia*, **66**, 558–562.

McInnes, P. F., 1989, Moose browsing and boreal forest vegetation dynamics, M.Sc. Thesis, University of Minnesota.

McInnes, P., Naiman, R. J., Pastor, J. and Cohen, Y., 1992, Effects of moose browsing on vegetation and litterfall of the boreal forest, Isle Royale, MI. *Ecology*, **73**, 2059–2075.

McNaughton, S. J., 1985, Ecology of a grazing ecosystem: the Serengeti, *Ecological Monographs*, **55**, 259–294.

Melillo, J. M., Aber, J. D. and Muratore, J. F., 1982, Nitrogen and lignin control of hardwood leaf litter decomposition dynamics, *Ecology*, **63**, 621–626.

Naiman, R. J., Johnston, C. A. and Kelley, J. C., 1988, Alteration of North American streams by beavers, *BioScience*, **38**, 753–762.

Northcott, T. H., 1971, Feeding habits of beaver in Newfoundland, *Oikos*, **22**, 407–410.

Orians, G. H. and Pearson, N. E., 1979, On the theory of central place foraging, in Horn, D. J., Stairs, G. R. and Mitchell, R. D. (Eds), *Analysis of Ecological Systems*, pp. 155–177, Columbus, OH: Ohio State University Press.

Pastor, J., Naiman, R. J. and Dewey, B., 1987, A hypothesis of the effects of moose and beaver foraging on soil nitrogen and carbon dynamics, Isle Royale, *Alces*, **23**, 107–124.

Pastor, J., Naiman, R. J., Dewey, B. and McInnes, P., 1988, Moose, microbes, and the boreal forest, *BioScience*, **38**, 770–777.

Peek, J. M., Ulrich, D. L. and Mackie, R. J., 1976, Moose habitat selection and relationships to forest management in northeastern Minnesota, *Wildlife Monographs*, **48**, 1–65.

Peterson, R. L., 1955, *North American Moose*, Toronto: University of Toronto Press.

Phillips, R. L., Berg, W. E. and Siniff, D. B., 1973, Moose movement patterns and range use in northwestern Minnesota, *Journal of Wildlife Management*, **37**, 266–278.

Pickett, S. T. A. and White, P. S. (Eds), 1985, *The Ecology of Natural Disturbance and Patch Dynamics*, Orlando, FL: Academic Press.

Rakestraw, L., Coffman, M. and Ferns, J., 1979, *Fire and Logging History of Voyageurs National Park*, Second Conference on Scientific Research in the National Parks, 26–30 November 1979, San Francisco, CA.

Reed, C., 1980, Habitat characterization: a production mode usage of a geographic information system, in Moore, P. A. (Ed.), *Computer Mapping of Natural Resources and the Environment*, Vol. 10, pp. 113–119, Cambridge, MA: Harvard Library of Computer Graphics, Harvard University.

Remillard, M. M., Gruendling, G. K. and Bogucki, D. J., 1987, Disturbance by beaver (*Castor canadensis* Kuhl) and increased landscape heterogeneity, in Turner, M. G. (Ed.), *Landscape Heterogeneity and Disturbance*, pp. 103–123, New York: Springer-Verlag.

Risenhoover, K. L., 1987, Winter foraging strategies of moose in subarctic and boreal forest habitats, Ph.D. Thesis, Michigan Technical Institute, Houghton.

Risenhoover, K. L. and Maas, S. A., 1987, The influence of moose on the composition and structure of Isle Royale forests, *Canadian Journal of Forest Research*, **17**, 357–366.

Rudersdorf, W. J., 1952, The coactions of beaver and moose on a joint food supply in the Buffalo River Meadows and surrounding area in Jackson Hole, Wyoming, M.Sc. Thesis, Utah State University, UT.

Senft, R. L., Coughenour, M. B., Bailey, D. W., Rittenhouse, L. R., Sala, O. E. and Swift, D. M., 1987, Large herbivore foraging and ecological hierarchies, *BioScience*, **37**, 789–799.

Smith, D. W. and Peterson, R. O., 1988, *The Effects of Regulated Lake Levels on Beaver at Voyageurs National Park, Minnesota* (US Department of Interior, National Park Service, Research/Resources Management Report MWR-11), Omaha, Nebraska: Midwest Regional Office.

Swanson, F. J., Kratz, T. K., Caine, N. and Woodmansee, R. G., 1988, Landform effects on ecosystem patterns and processes, *BioScience*, **38**, 92–98.

Telfer, E. S., 1984, Circumpolar distribution and habitat requirements of moose (*Alces alces*), in Olson, R., Hastings, R. and Geddes, F. (Eds), *Northern Ecology* and *Resource Management*, pp. 145–182, Edmonton, Alberta: University of Alberta Press.

Tiedemann, A. R. and Berndt, W. H., 1972, Vegetation and soils of a 30-year deer and elk exclosure in central Washington, *Northwest Science*, **46**, 59–66.

Tomlin, C. D., Berwick, S. H. and Tomlin, S. M., 1980, The use of computer graphics in deer habitat evaluation, in Moore, P. A. (Ed.), *Computer Mapping of Natural Resources and the Environment*, Vol. 10, pp. 125–131, Cambridge, MA: Harvard Library of Computer Graphics, Harvard University.

Van Horne, B., 1983, Density as a misleading indicator of habitat quality, *Journal of Wildlife Management*, **47**, 893–901.

Voight, G. K. and Steucek, G. L., 1969, Nitrogen distribution and accretion in an alder ecosystem, *Soil Science Society of America Proceedings*, **33**, 946–949.

Wolfe, M. L., 1974, An overview of moose coactions with other animals, *Naturaliste Canadien*, **101**, 437–456.

18

The ecological interpretation of satellite imagery with special reference to bird habitats

G. H. Griffiths, J. M. Smith, N. Veitch and R. Aspinall

Introduction

An understanding of the habitat preferences of a species is important, not only to improve ecological knowledge about that species but also to devise strategies for the protection of its habitat. In this respect, there is a requirement to develop methodologies for the rapid and cost-effective mapping and monitoring of extensive areas of land to determine their conservation value. This is especially true in the uplands of Britain where rapid land use change, resulting from afforestation and agricultural improvement (RSPB, 1986), has put pressure on remaining semi-natural habitats and results in an urgent need for their conservation.

Although a considerable amount of qualitative information exists which describes the habitat preferences of various species, such studies are frequently localized in nature. To some extent this is the result of the difficulties inherent in collecting habitat information by conventional methods, including air-photo interpretation and field survey, and quantifying these data for statistical analysis.

The possibility of classifying extensive areas of land very rapidly from digital satellite imagery provides, for certain types of landscapes and species, increased opportunities to develop quantitative models of the relationship between land cover and species diversity and abundance. The very large number of samples that can be extracted from satellite imagery provides a rigorous basis for the development and testing of statistical ecological models. Results of such models can subsequently be applied to the broad areas covered by satellite images (e.g. SPOT satellite images with a scene coverage of 60 km × 60 km and Landsat images with a scene coverage of 185 km × 185 km).

This chapter presents the preliminary results from two case studies in which census data of upland bird species have been used in conjunction with vegetation ground survey and satellite data to establish and test models of the relationships between land cover and spatial pattern and selected upland bird species. Two analytical approaches are used based on:

1. relating species abundance to habitat data using regression; and
2. calculating the probability of presence of a species using an inductive learning procedure for pattern recognition based on Bayes' theorem.

Habitat preferences

Land cover

Land cover is clearly an important determinant of species abundance and diversity, with different species depending upon different vegetation communities for cover and food. A number of studies have demonstrated the importance of land cover as a major determinant of species type and number.

Bibby (1986) quantified the very close association between the presence of heather moorland and nest occupancy of merlins (*Falco columbarius*) in Wales. Major vegetation communities were identified at a distance of up to 4 km from each nest site from maps produced by the Second Land Utilization Survey. Discriminant analysis was used to isolate which of five vegetation types (heather, bracken, grass moorland, conifers and farmland) in three distance ranges (0–1, 1–2 and 2–4 km) predicted the most and least occupied sites between 1970 and 1984. The results showed that occupied sites on average had greater heather and bracken cover and less grass moorland at all ranges, with the model able to predict the correct level of occupancy with over 90% accuracy. Although the model was able to quantify a previously known preference of merlin for heather moorland, Bibby (1986) concludes that it is difficult to provide an ecological interpretation for this preference. In this sense quantitative ecological models may provide insights into habitat selection behaviour but may not be capable of explaining such preferences without additional ecological information about the species in question.

Ratcliffe (1976) in a study of breeding golden plover (*Pluvialis apricaria*) in Britain provides data which show that high to very high breeding densities (>4–10 pairs/km²) are associated with the presence of substantial areas of limestone grassland which have developed on base-rich soils; conversely, low density (<4 pairs/ km²) is characteristic of more acid grassland and mountain bog. These differences in density can be explained by variations in soil fertility; invertebrate populations are much higher on limestone and alluvial grassland than on blanket bog with highly acidic and base-poor peat. In this respect land cover is being used as a surrogate for soil fertility which influences species number.

Newton (1986) in a study of the nest spacing of sparrowhawks (*Accipiter nisus*) discovered that nest spacing varied in different regions of Britain. Nest spacing was found to be highly correlated with altitude; nests being further apart at higher altitudes. Nest spacing was also correlated with land productivity as measured from an agricultural map, spacing increasing as land fertility declines. Although Newton (1986) stated that the relationship between nest-spacing and landscape was useful, because it enables predictions of maximum sparrowhawk densities to be made from simple map data, it is unlikely that sparrowhawks responded to elevation and land productivity as such. In fact nests are found to be furthest apart in areas of low prey density; elevation and land productivity are simple surrogate variables for prey density.

Landscape pattern

The pattern of various elements within the landscape, in addition to the type and area of land cover features, has also been demonstrated to be an important determinant of species diversity and abundance.

Forman and Godron (1986) conceptualize pattern in terms of patches (communities or species assemblages) surrounded by a matrix with a dissimilar community structure or composition. Pattern in the landscape can be measured at a number of levels, but two in particular are fundamental; the measurement of frequencies of object (patch) characteristics, (e.g. numbers of patches in a specific size class, diversity of patch types) and the spatial relationship between different objects, (e.g. inter-patch distance). Spatial pattern occurs at all landscape scales; in ecological terms however, the critical scale for analysis is related to the scale of species behaviour, e.g. territory, home range, breeding dispersion, etc.

It is neither desirable nor possible to produce a single index of pattern, but a number of researchers have studied particular elements of pattern in relation to species distributions. The overall objective of these studies has been to predict quantitatively the distribution of a species from the spatial arrangement of a set of habitat patches and the landscape structure of the surrounding region.

Patch size and isolation

Much of the research on landscape pattern related to species abundance and diversity has been based on the concept of the species–area curve, which describes the relationship between the number of species in a particular habitat and the size of the habitat (Greig-Smith, 1983).

This concept is central to the theory of island biogeography (MacArthur and Wilson, 1967) based on studies of species numbers recorded for islands of different size and isolation. The theory attempts to explain the number of species on an island as a function of island area, isolation and age. From this work similar relationships were discovered to apply to terrestrial 'islands'; in particular the number of bird species in woods of different size and internal heterogeneity have been extensively studied (Moore and Hooper, 1975; Galli *et al.*, 1976). For example, the effect of woodland area and isolation on forest bird communities in the Netherlands is strongly associated with wood size and negatively with isolation. In this study the isolation of a wood was defined as the distance between the wood and the nearest extensive forest containing a pool of potential colonizing species.

Alternatively, isolation can be related to distance from a food supply. Bibby (1986) demonstrated that a clutch size of five for breeding merlins was correlated with proximity to farmland within a radius of 1–4 km of the nest site. Merlin rarely lay other than four or five eggs in a clutch; the higher clutch size in nests closer to productive farmlands may be explained by the higher density of small bird prey species on farmland in contrast to the remoter moorland, particularly in early Spring.

Habitat heterogeneity

It is difficult to separate the effects of the habitat heterogeneity within a patch from the effects of patch size; the occurrence of many species can be predicted better from structural and floristic characteristics of woods than from patch size and isolation. It would appear that both patch size and habitat diversity are important factors determining species number; their relative effects will vary with the species being considered and the scale of analysis. Information on both can be derived from analysis of satellite images.

Satellite studies of wildlife habitats

Most studies of the effects of landscape pattern have investigated relationships between woodland bird numbers and the size, composition and isolation of woods in lowland agricultural landscapes. Landscapes of this type, which can be considered as relatively simple systems with woods representing terrestrial islands in a relatively uniform matrix of agricultural land, provide more easily described experimental locations than upland landscapes. In addition, species number and diversity are more easily recorded in discrete woodland blocks and habitat data (wood size, internal habitat heterogeneity and isolation) can be derived relatively easily from field survey, map information and air-photo interpretation. For larger areas, conventional methods of data collection, both for wildlife and habitat data, become limiting and alternative techniques may need to be used. There are a number of examples of the application of satellite imagery to classify land cover for wildlife habitat mapping, particular from North America, some of which have incorporated elements of pattern in the analysis.

In a study of bird distribution in relation to the area and distribution of xeric and mesic forest, farmland and rangeland, Palmeirim (1988) used classified Landsat TM data to produce probability maps depicting the likelihood of occurrence of selected species. This study relied heavily on existing knowledge of the habitat requirements of a species to incorporate spatial factors into the analysis. For example, it was known that the red-eyed vireo (*Vireo olivaceus*) avoids the edges of forests. The Landsat derived probability map was reprocessed using geographic information system (GIS) software to show a decreased probablity of finding this species near to the edge of mapped woodland areas. The ornithological literature also suggested that, because of this species' reluctance to fly in open spaces, the whole territory of the red-eyed vireo must be within a continuous forest canopy. Since the territory of this species is about 1 ha, all woodlots smaller than 1 ha were eliminated from the habitat suitability map. Habitat suitability maps for different species can thus be generated by using a GIS system in conjunction with sets of rules about habitat preferences. These maps can depict areas of high species abundance and diversity or relative ecological value for a species.

Avery and Haines-Young (1990) used Landsat MSS imagery to predict dunlin (*Calidris alpina*) numbers in the Flow country of northern Scotland. The near infra-red band 7 of Landsat MSS in addition to its sensitivity to green vegetation, is also sensitive to soil wetness. The habitat favoured by dunlin is wet moorland interspered by small pools, so that dunlin numbers are significantly negatively correlated with an index of soil wetness derived from the MSS band 7 reflectance values. This relationship was used to predict dunlin numbers on a random sample of unsurveyed sites and hence to assess the damage to habitat from afforestation over time.

The application of habitat rules to derive and present ecological maps has merit, but the digital nature of thematic land cover maps derived from classified satellite imagery is well suited to the analysis of spatial pattern and relationships between wildlife and habitat. The techniques for the analysis of spatial relationships for raster images are well developed (Joyce-Loebel, 1985) and a number of researchers have applied them to classified image data (Janssens and Guilink, 1988; Griffiths and Wooding, 1989).

To illustrate the potential, samples of classified Landsat TM data for a region covering the southern Uplands/northern Pennines were analysed to derive indices of land-cover diversity and patch size. The results are presented according to the ITE Land Classification System which stratifies Great Britain into 32 land classes according to a range of environmental variables. By presenting the results according to the Land Classification System, land cover diversity (i.e. the number of different land cover types in each 1 km × 1 km square (Figure 18.1) and patch size for heather (i.e. the area of mature heather in each 3 km × 3 km sample area within different patch size ranges) (Figure 18.2) can be compared for different landscape types in the study areas (land classes 19, 20 and 22). The results show the very marked difference in the number of cover types in each 1 km × 1 km square (Figure 18.1), particularly between land class 19 which is relatively heterogeneous with a

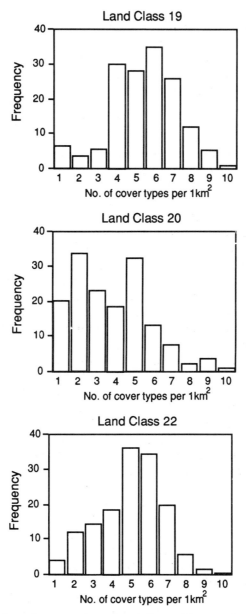

Figure 18.1. The number of cover types within 1 km × 1 km sample square in the southern Uplands/northern Pennines study area plotted against the frequency of occurrence in land classes 19, 20 and 22.

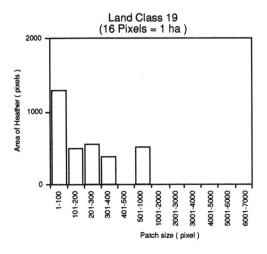

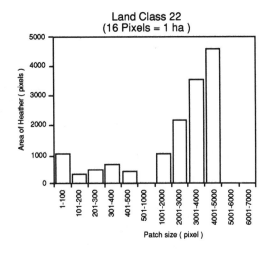

Figure 18.2. The area of heather moorland plotted against patch size derived from 28 3 km × 3 km sample squares in the southern Uplands/northern Pennines study area.

large proportion of samples having a high number of cover types and land class 20 which is more uniform with a smaller proportion of samples containing a large number of cover types. Figure 18.2 shows a marked contrast between land class 22, which contains a few large patches characteristic of extensive moorland, compared with land class 19, which contains a relatively large number of small patches suggesting a more fragmented landscape.

Analysis of satellite imagery

Land cover

Two separate studies have recently been undertaken to explore the potential of satellite imagery for mapping upland bird habitats. The first of these studies is a collaborative project between the National Remote Sensing Centre (NRSC) (now the National Remote Sensing Centre, Limited) and the Royal Society for the Protection of Birds (RSPB) to study selected upland bird species in the northern Pennines and southern Uplands. The second study, also on upland birds, has been undertaken by the Macaulay Land Use Research Institute (MLURI) in conjunction with the Nature Conservancy Council (NCC) (now Scottish Nature) for a study area in Grampian Region.

For the NRSC/RSPB study the Institute of Terrestrial Ecology (ITE) Land Classification System (Bunce *et al.*, 1981) was used to select a random sample of 32 1 km × 1 km squares in the northern Pennines and southern Uplands (Figure 18.3). The ITE Land Classification System (Bunce, 1983) is being used increasingly to provide a framework for environmental monitoring because of its demonstrated effectiveness for sampling variability in the landscape. In this study the Land Classification System was used to select sample squares centred on 1 km × 1 km squares classified as upland land classes.

The 32 1 km × 1 km squares, arranged in 16 pairs approximately 5 km apart, form the centre of 3 km × 3 km blocks in which field survey was used to map major

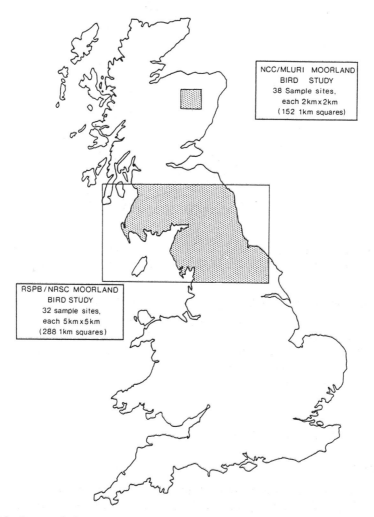

Figure 18.3. Case study locations.

vegetation units and to collect bird census data. This gives a total of 288 1 km ×
1 km squares. The majority of the 1 km × 1 km squares (more than 75%) are in ITE
Land Classes 19 (smooth hills, mainly heather moor, often afforested), 20 (mid-valley
slopes with a wide range of vegetation types) and 22 (margins of high mountains,
moorland, often afforested).

The study area is covered by three separate Landsat thematic mapper (TM)
scenes. Two scenes were acquired for 17 April 1987 and one for 14 May 1988. Two
separate classifications were developed for these scenes.

The imagery was geometrically transformed to the UK national grid and image
data covering 5 km × 5 km sample areas centred on 3 km × 3 km census sites
extracted from each corrected scene. Within each extract, visual interpretation
methods were used to produce a mask of lowland areas, and a mask of upland,
semi-natural vegetation. The distinction between these two areas is clearly visible on
the imagery. A preliminary analysis of the image data showed that certain features,
including broadleaf woodland and upland bog were mapped more effectively using
visual interpretation of imagery where context and pattern assist in the analysis. The
effective discrimination of broadleaf woodland for example, would have required
multi-seasonal imagery, which was not available for 1988 in the project area.

Land cover features which were spectrally separable (e.g. water, coniferous
woodland and mature and pioneer heather) were classified automatically by supervised maximum likelihood classification using four spectral channels; the visible red
(TM3), the near infra-red (TM4) and two in the middle infra-red (TM5 and TM7). In
the remaining areas of upland semi-natural vegetation with a high proportion of
mixed pixels (i.e. pixels containing more than one cover type), an unsupervised clas-

Table 18.1. *Vegetation classes classified from the Landsat TM imagery in*
northern England/southern Uplands, with abbreviations.

	Variable name*

Woodland:
    coniferous,
    broadleaf and mixed	

Lowland agricultural: arable crops, improved and permanent pasture	
Semi-natural vegetation: heather (mature),	UP1
heather (pioneer),	UP2
heather/moor grass mixed (heather dominant),	UP3
moor grass/heather mixed (moorgrass dominant),	UP4
moor grass (1),	UP5
moor grass (2),	UP6
upland pasture/moor grass mixed,	UP7
moor grass/upland pasture mixed,	UP8
upland pasture,	UP9
bracken/juncus	UP10
Bog	Bog
Water	W
Shadow	

* See text for explanation of variables used in analysis.

sification technique was employed. This technique divides the multispectral image data into separate spectral regions statistically. These can be amalgamated later and interpreted with reference to ground data information.

Combining the classification of the lowland and upland masks produced the list of classes shown in Table 18.1. The percentages of each land-cover type within each 1 km × 1 km square for 32 3 km × 3 km sample areas were derived from the final image classifications.

For the MLURI study the geographical spread of sample sites is more restricted (Figure 18.3) with 38 tetrads (2 km × 2 km) within a study area of approximately 40 km × 40 km to give a total sample of 152 1 km × 1 km squares. A Landsat TM image from 17 April 1987 was geometrically corrected and bands 3, 4 and 5 were used for classification.

A similar hierarchical system for classification was adopted (Figure 18.4) with the aim of characterizing variation in grassland and moorland areas; other features, including forest, snow and water were excluded from the classification using visual interpretation to identify a series of masks. Remaining moorland and grassland areas were separated using supervised maximum likelihood classification applied to the image after smoothing with a 10 × 10 filter. This smoothing process, which removes high frequency image components, improved discrimination between moorland and grassland classes. The moorland and grassland classes were each classified into 10 spectral classes using an unsupervised minimum distance to means algorithm.

Field survey; vegetation mapping and bird census data

For the NRSC/RSPB study, each 3 km × 3 km sample square was visited twice between April and June 1988 and the number of breeding birds recorded according to standard census techniques. At the same time the recorder mapped the major vegetation communities at 1 : 10 000 scale using a field recording technique developed by ITE for the Land Classification System. These data were used during classification of the satellite imagery to derive training statistics and to label classes. This process was assisted by producing hard-copy photographic imagery of the sample squares and relating colours on the image directly to vegetation types in the field by NRSC staff.

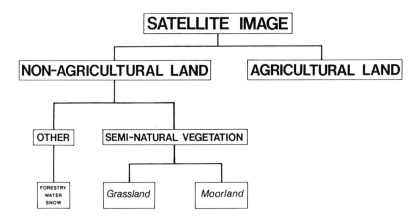

Figure 18.4. Hierarchical image segmentation procedure.

Table 18.2. Moorland spectral classes 1, 5, 7 and 10 described according to plant communities.

Spectral classes	Moor 1	Moor 5	Moor 5	Moor 10
Boreal heather moor	64.0	23.1		67.7
Alpine azalea–lichen heath	13.0	20.6		5.5
Mountain blanket bog	11.4	14.7		
Blaeberry heath		12.5	10.4	
Blanket bog		10.3	13.1	
Bog moss water track			25.8	
White bent–tussock grass			11.3	
Heath rush–fescue grass			8.6	
Deer–grass moor			5.7	
Atlantic heather moor			8.8	7.9
Conifer forest				14.2

* After Robertson (1984).

In the MLURI study the spectral classes are described according to plant community types mapped for selected areas during field survey. Although there are no direct and exclusive relationships between spectral classes and individual plant communities, particular communities are more frequently represented in some spectral classes than others. This is summarized in Table 18.2 for four of the moor classes derived from the unsupervised classifier and compared with field survey data. For example, 64% of spectral class Moor 1 was mapped as Boreal Heather Moor during field survey with 13% Alpine azalea–lichen heath and a further 11.4% mountain blanket bog. Bird census data were collected using standard NCC moorland bird survey methods.

Landscape pattern

In both studies a number of pattern measures as well as other environmental variables were derived from image classifications and included in the analysis of bird–habitat relationships. The pattern measures derived included:

1. number of patches of mature and pioneer heather,
2. distance from agricultural land, and
3. altitude and slope.

The number of patches of mature heather in one of the 3 km × 3 km sample areas in the southern Uplands/northern Pennines study area are illustrated in Figure 18.5. The larger the size of patch the brighter the grey tone on the black and white image.

In both studies, distance to agricultural land was calculated using proximity software which assigns progressively higher values (digital numbers) to pixels at greater distance from agricultural land. An example is illustrated in Figure 18.6 in which the distance of each pixel from agricultural land is displayed as a grey tone. Average altitude and slope data for each 1 km × 1 km sample area were derived from a 100 m grid cell digital elevation model of the UK.

Development of ecological models

Two very different approaches were used to model the relationships between the habitat data derived from the classified satellite imagery and bird census data.

Figure 18.5. *Varying patch sizes of mature heather displayed in grey tones for a 5 km × 5 km sample square in the southern Uplands/northern Pennines study area.*

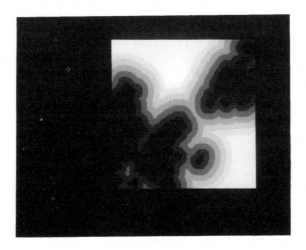

Figure 18.6. *Proximity image displaying distance to agricultural land for a 5 km × 5 km sample area in the southern Uplands/northern Pennines study area.*

Multiple regression

In the RSPB study step-up multiple regression was employed to select the optimum set of predictor variables, i.e. the fewest number of variables to explain as much of the observed variation in bird numbers as possible. The strategy adopted known as forward selection procedure, was to select a single variable and then add variables one at a time until the addition of further variables failed to produce an appreciable increase in the coefficient of determination (R^2). By calculating partial correlations for all possible candidate variables, the candidate variable with the strongest partial correlation is the predictor to be added.

This analysis was carried out by the RSPB using the SYSTAT statistical package. The original approach was to use the percentage of different satellite derived cover classes within each 1 km² as the predictor variables. An analysis of variance showed that statistical differences occurred between the dispersed sample of 3 km × 3 km sample areas but, because of spatial autocorrelation, not between individual 1 km × 1 km squares within each sample area. For this reason the analysis used the mean of the percent cover of each cover type within each 3 km × 3 km square regressed against the mean of the observed number of bird species. Bird species modelled are red grouse (*Lagopus lagopus*), golden plover, curlew (*Numenius arquata*) and lapwing (*Vanellus vanellus*) in the 3 km × 3 km area.

Preliminary results (Table 18.3) are presented for three levels of analysis: (a) land cover alone, (b) land cover and proximity to agricultural land, and (c) land cover, proximity to agricultural land and number of patches of mature heather (see above).

This preliminary analysis suggests that land cover alone, i.e. the proportion of each cover type in 3 km × 3 km is an important determinant of red grouse (81.4%) and golden plover (64.2%), but less effective as an indicator of curlew (45.9%) and lapwing numbers (30.0%). Both red grouse and golden plover are associated with a heather/grass moorland association (UP3), though UP2, pioneer heather, is involved in explaining an additional 10% of the numbers for this species.

Proximity of agricultural land (b) does not affect the predictions for red grouse or golden plover but appears to be negatively related to the presence of curlew, i.e. where agricultural land is more than 1 km away, fewer curlew occur. Proximity to agricultural land is positively related to lapwing numbers, i.e. where agricultural land is closer than 500 m more lapwing are present.

The number of mature heather patches in each 3 km × 3 km sample area appears to be significant for golden plover, explaining an additional 10% of the

Table 18.3. Habitat variables involved in explaining (percent) variations in distribution of red grouse, golden plover, curlew and lapwing at three levels of analysis: (a) land cover proportions; (b) land cover and distance to agricultural land where D1 is <500 m, D2 is 500–1000 m and D3 is >1000 m; and (c) land cover, distance to agricultural land and number of patches of mature heather where NOUP1 is number of mature heather patches and NOUP2 is number of pioneer heather patches. *

Species	Variable (a)	%†	Variable (b)	%†	Variable (c)	%†
Red grouse	UP3	72.6	UP3	72.6	UP3	72.6
	W	78.1	W	78.1	W	78.1
	UP2	81.4	UP2	81.4	UP2	81.4
Golden plover	UP3	64.2	UP3	64.2	UP3	64.2
					NOUP1	74.5
Curlew	UP4	12.3	UP4	12.3	NOUP2	12.4
	W	24.8	−D3	30.7	−D3	22.7
	UP6	37.0			UP4	38.3
	Bog	45.9			−W	46.0
Lapwing	UP2	20.0	UP2	20.0	NOUP2	36.7
	UP5	30.0	UP5	30.0	−D3	46.2
			D1	37.7		

* For a description of the other variables see Table 18.1.
† Cumulative percent variance explained.

variance. In the case of lapwing, a combination of the number of patches of pioneer heather and distance from agricultural land explains 46.3% of the variance; the area of land cover is not as important a determinant in the prediction of this species.

Model testing

The model was tested by applying the regression equations developed from the 1988 data to a new and independent bird census data set collected in a similar fashion to the 1988 bird census data, but for a different set of sample sites in northern England and the southern Uplands. The results, which are summarized in Table 18.4, were disappointing, with only the prediction for red grouse being statistically significant.

One possible reason for the poor predictive power of the model based on the satellite classification is that the relationship between bird numbers and land cover apparent in the 1988 data set occurred entirely by statistical chance and that no consistent relationships exist between species abundance and the habitat classes classified from the imagery.

It is possible that factors other than vegetation measures are more important determinants of species abundance. Ratcliffe (1976) in a study of breeding golden plover in Britain showed that high breeding densities are associated with base-rich soils where vegetation supports higher invertebrate populations. Fuller (1982) reports that although red grouse are strongly dependent upon heather shoots for feeding, the relationship between grouse numbers and heather abundance is not simple as the underlying geology affects the nutritional value of the heather. Perhaps the vegetation characteristics which influence species abundance most are not accounted for by the image classification or landscape pattern parameters. For example, vegetation wetness is known to be an important determinant of the distribution of certain upland bird types, but wetness is extremely difficult to infer from the satellite imagery.

A statistical comparison of bird numbers against ITE land class for both the 1988 and 1989 data sets showed that the numbers of many bird species vary significantly between ITE land classes, suggesting that the ITE Land Classification System may contain information about variables which influence bird numbers and which are not contained in the satellite classifications.

The fact that the relationships observed in the 1988 data set could be explained by known bird habitat preferences and that relationships within the 1989 data set

Table 18.4. Test of predictions of bird numbers for sites surveyed in 1989.

Species	% of variance explained	p
Red grouse (RG)	42	—*
Golden plover (GP)	6	NS
Curlew (CU)	1	NS
Stonechat (S)	1	NS
Meadow pipit (MP)	0	NS
Oystercatcher (OC)	(29)	(**)
Whinchat (WC)	12	NS
Ring ouzel (RZ)	7	NS
Snipe (SN)	1	NS
Lapwing (L)	0	NS

NS, not significant; * $p < 0.001$.

were weaker suggests that the 1988 classifications may have classified bird habitat variables more successfully than the 1989 classifications.

Probabilistic modelling

A different modelling procedure was employed for the study in the Grampian region. In this case the distribution is to be predicted (rather than abundance). The data are analysed using a procedure for pattern recognition (Williams *et al.*, 1977; Grubb, 1988) based on Bayes' theorem allowing relationships between bird distribution and mapped habitat information to be learned. The major attraction and one of the advantages of this pattern recognition procedure is that it emulates the intuitive processes by which relative quality of habitats are assessed by biologists and quantifies important habitat characteristics.

Habitat suitability of a site is defined by presence or absence of a species. The frequency of specified habitat characteristics, e.g. area of cover type, altitude, distance from agricultural land, is then calculated for each habitat suitability class and used to calculate conditional probabilities. The significance of conditional probabilities for discriminating between habitat suitability classes is determined with χ^2. Conditional probabilities are then combined into a single *a posterior* probability describing habitat suitability through use of Bayes' theorem. Thus the probability of suitable conditions is calculated from known habitat classes and then used to predict a habitat suitability class for sites where habitat conditions are known but where suitability is unknown.

The model presented here combines topographic, locational and habitat (spectral) information summarized for 1 km × 1 km square sample sites. No prior assumptions about sample squares are made and all prior probabilities are set at 0.5. This prior probability is modified using Bayes' theorem and conditional probabilities for habitat and topographic variables.

Programs to implement pattern recognition were written within the GENSTAT statistical package (GENSTAT 5 Committee, 1987) on a VAX 3600 minicomputer at the Macaulay Land Use Research Institute. The programmes are linked to an ERDAS image processing system housed on a Wyse microcomputer through DECNET. Each topographic, locational and habitat variable is searched for a value which optimizes the discriminatory difference between habitat suitability classes; this is measured through χ^2 analysis of frequencies.

Models for the distribution of curlew and golden plover were developed from analysis of bird distribution data collected in 1988. Eight habitat characteristics were identified as significantly associated with presence/absence for golden plover and sixteen for curlew.

The conditional probabilities for golden plover being present are higher than probabilities for absence in 1 km × 1 km squares with minimum altitude above 600 m, maximum altitude greater than 825 m and in squares with minimum distance from agricultural land greater than 1.8 km. For spectral class information classes, conditional probabilities for absence are greater than those for presence; squares where six or more of the ten moorland spectral classes are present and squares with a total of 58% moorland have high probability of golden plover being present. Squares containing more than 28% of moor 1 and moor 5 (Table 18.2) in a square also show a greater probability of golden plover being present. Thus golden plover are more likely to be present at high altitude and in areas where there are few extensive patches of heather dominated vegetation. This is in broad agreement with known golden plover habitat preferences (Ratcliffe, 1976) and may be related to the preference of this species to avoid areas of clumpy, mature heather which reduces visibility across its territory.

For curlew, the conditional probabilities for absence are greater than those for presence for all topographical and locational characteristics. Minimum altitude above 500 m, maximum altitude above 700 m, minimum distance to agricultural land greater than 3.4 km and maximum distance greater than 4.4 km reduce the probability of curlew being present. The total proportion of grassland within squares is also significant, with higher probability for presence than absence. Individual grassland classes also have a weak effect, again the critical proportions suggest 'presence' of grassland to be significant. The presence of moor classes 1, 6 and 10 (Table 18.2) reduces the probability of curlew being present. Moor class 7, which is a grassland plant community, has the opposite effect; more than 4% of the class in a square producing a conditional probability in favour of presence.

Model development
Four habitat characteristics were used in the model for golden plover and for curlew (Table 18.5). For golden plover these were: minimum altitude, number of moorland types and moor spectral classes 1 and 5. The overall accuracy of the model for golden plover is 57.2% (1988 data). For curlew the habitat characteristics used were minimum altitude, number of grassland types and moor spectral classes 7 and 10. The model for curlew has an overall accuracy of 80.3%.

Model testing
The models were tested against bird survey data collected during 1989 and the results are presented in Table 18.6. Accuracy of prediction is similar to that achieved during model development.

Discussion

The results from the two case studies presented, suggest that probabilistic modelling based on habitat suitability criteria, is a successful method for predicting the dis-

Table 18.5. *Model development for golden plover and curlew.*

	Observations			
	Absent	Present	Row total	Row %
Golden plover				
Model Absent	38	42	80	47.5
Prediction Present	23	49	72	68.1
Column total	61	91	152	
Column %	62.3	53.8	Overall accuracy 57.2%	
Curlew				
Model Absent	51	5	56	91
Prediction Present	25	71	96	74
Column total	76	76	152	
Column %	67.1	93.4	Overall accuracy 80.3%	

Table 18.6. Model testing (1989 bird census data)—golden plover and curlew.

| | Observations | | | |
	Absent	Present	Row total	Row %
Golden plover				
Model Absent	12	16	28	42.9
Prediction Present	15	39	54	72
Column total	27	55	82	
Column %	44.4	70.9	Overall accuracy 62.2%	
Curlew				
Model Absent	53	8	61	86.9
Prediction Present	11	10	21	47.6
Column total	64	18	82	
Column %	82.8	55.6	Overall accuracy 76.8%	

tribution (presence/absence) of selected upland bird species. The alternative method based on regression analysis for the prediction of bird abundance, showed significant correlations between land cover and bird numbers for the 1988 data set, but testing of the model on an independent data set for 1989 was less successful. However, it would be premature to abandon this technique until further research is undertaken to identify exactly why the model failed to predict bird numbers on this independent data set.

In terms of providing a means for achieving rapid and cost-effective mapping of extensive areas, the results of both case studies are sufficiently encouraging to undertake further development and testing. Although the analytical methods employed for relating bird data to habitat information are very different between the two studies, both possess similar advantages. In particular, both methods assess likely distribution/abundance on consistent criteria across the study area and use quantified criteria against which habitat is assessed and predictions made. These advantages provide the methods with great potential for application in conservation evaluation, allowing objective assessments of habitat to be made.

The application of satellite imagery to the evaluation of upland bird habitat offers a rapid means of data collection; when linked to modelling of the type described in these two case studies these data can provide a valuable means of assessing habitat and distribution of upland bird species. Appropriate analysis of imagery has the advantage of providing relevant and useful information on habitat over extensive areas, such data being unavailable through more conventional habitat survey methods. Although data derived from imagery may not match existing habitat classifications or ecological information, the benefits which can be gained by investigation of the variety of habitat measures this source suggest that the methodologies and applications described have much to offer ecological study.

Acknowledgements

The MLURI study was funded by the Nature Conservancy Council and the RSPB study by the Directorate of Rural Affairs, Department of Environment. Statistical analysis of the data from the southern Uplands/northern Pennines study was conducted by Mark Avery at the RSPB.

References

Avery, M. I. and Haines-Young, R. H., 1990, Population estimates for the dunlin (*Calidris alpina*) derived from remotely sensed satellite imagery of the Flow Country of northern Scotland, *Nature*, **344**(6269), 860–862.

Bibby, C. J., 1986, Merlins in Wales: site occupancy and breeding in relation to vegetation, *Journal of Applied Ecology*, **23**, 1–12.

Bunce, R. G. H., Barr, C. J. and Whittaker, H. A., 1981, A stratification system for ecological sampling, in Fuller, R. M. (Ed.), *Ecological Mapping from Ground, Air and Space*, ITE Symposium No. 10, Monkswood, 25–27 November, 1981, pp. 39–46.

Forman, R. T. T. and Godron, M., 1986, *Landscape Ecology*, New York: Wiley.

Fuller, R. J., 1982, *Bird Habitats in Britain*, Calton: T. & A. D. Poyser.

Galli, A. E., Leck, C. F. and Forman, R. T. T., 1976, Avian distribution patterns in forest islands of different sizes in central New Jersey, *The Auk*, **93**, 356–364.

GENSTAT 5 Committee, 1987, *GENSTAT 5 Reference Manual*, p. 749, Oxford: Clarendon.

Greig-Smith, P., 1983, *Quantitative Plant Ecology* (*Studies in Ecology, Vol. 9*), 3rd Edn., Oxford: Blackwell Scientific.

Griffiths, G. H. and Wooding, M. G., 1989, *Use of Satellite Data for the Preparation of Land Cover Maps and Statistics, Vol. II. Ecological Consequences of Land Use Change (Final Report to the Department of the Environment)*, p. 95.

Grubb, T. G., 1988, *Pattern Recognition—a Simple Model for Evaluating Wildlife Habitat, Research Note RM-487*, Rocky Mountain Forest and Range Experiment Station, USDA Forest Service.

Janssens, P. and Gulinck, H., 1988, Connectivity, proximity and contiguity in the landscape; interpretation of remote sensing data, in *Connectivity in Landscape Ecology; Proc. 2nd International Seminar on the International Association for Landscape Ecology*, Munster, 1988.

Joyce-Loebel (Ed.), 1985, *Image Analysis: Principles and Practice*, p. 250.

MacArthur, R. H. and Wilson, E. D., 1967, *The Theory of Island Biogeography* (*Monographs in Population Biology 1*), Princeton, NJ: Princeton University Press.

Moore, N. W. and Hooper, M. D., 1975, On the number of bird species in British woods, *Biological Conservation*, **8**, 239–250.

Newton, I., 1986, *The Sparrowhawk*, Calton: T. & A. D. Poyser.

Palmeirim, J. M., 1988, Automatic mapping of avian species habitat using satellite imagery, *OIKOS*, **52**, 59–68.

Ratcliffe, D. A., 1986, Observations on the breeding golden plover in Great Britain, *Bird Study*, **23**, 63–116.

Robertson, J. S., 1984, *A Key to the Common Plant Communities of Scotland (Soil Survey of Scotland Monograph)*, Aberdeen: The Macaulay Institute for Soil Research.

RSPB, 1986, *Forestry in the Flow Country—The Threat to Birds. A Critique of Afforestation in East Sutherland and Caithness*, p. 7. RSPB.

Williams, G. L., Russell, K. R. and Seitz, W. K., 1977, Pattern recognition as a tool in the ecological analysis of habitat, in *Classification Inventory and Analysis of Fish and Wildlife Habitat: Proceedings of a National Symposium*, Phoenix, Arizona, pp. 521–531, US Fish and Wildlife Service, Office of Biological Service.

19

The use of landscape models for the prediction of the environmental impact of forestry

C. Lavers and R. Haines-Young

Introduction

The Flow Country of Caithness and Sutherland in Scotland is an area of international significance for conservation. It holds some of the largest populations of wading birds in the European Community. Species include dunlin (*Calidris alpina*), golden plover (*Pluvialis apricaria*) and greenshank (*Tringa nebularia*). Despite its ecological value, however, it has been under development pressure. In the late 1970s and most of the next decade, there was a large amount of afforestation in the area, which many felt was undermining its ecological value. As a result, forestry in the Flow Country became the focus of an intense public debate (Bainbridge *et al.*, 1987).

At the end of the 1980s forestry policy in the Flow Country was under review. This chapter describes how the environmental impact of this policy was evaluated by a joint project with the Royal Society for the Protection of Birds (RSPB), one of the 'key players' concerned with forestry policy in the area. In particular the chapter illustrates how models of landscape structure can be important in making such 'environmental audits'.

The Flow Country landscape

The Flow Country landscape is simple and highly distinctive. The area is an extensive peat covered plateau. The areas favoured by dunlin and golden plover consist of a tight jigsaw of arcuate or irregularly shaped peat lined pools cut into the vegetation matrix. The pools, known locally as dubh lochans, are small, generally less than 100 m across. They occur in clusters or complexes, rarely extending for more than 1 km in any direction. Field observations of dunlin and golden plover, in particular, suggest that their distribution may be controlled by the presence of these habitat patches. Dunlin especially, are not often seen far from dubh lochan systems.

The goal of the initial phase of the project was to attempt to map the distribution and abundance of wading birds in the Flow Country. Remotely sensed satellite information was used to characterize variations in landscape structure of the area and to predict habitat quality and population numbers. A Landsat MSS image for

May 1978 was used in the study because it was one of the only cloud-free images of the area. It also pre-dated most of the afforestation. Despite its coarse resolution it was considered that the sub-pixel pool features would modify the spectral signature of the peat complexes sufficiently to make their presence detectable.

The initial study

The ecological model developed in the initial study was a very simple one. Since it has been reported elsewhere it will be described only briefly here (Avery and Haines-Young, 1990).

The key assumptions of the model were that the habitat favoured by wading birds are the dubh lochan systems, where tussock peat is interspersed with small pools. Areas which are avoided are large water bodies, woodland and agricultural land. They also tend to avoid steeply sloping sites and altitudes over 550 m. The near infra-red band 7 of Landsat MSS is known to be sensitive to variations in vegetation performance and ground wetness. Thus by using a mask to remove all the areas in the Flow Country which would be unsuitable for wading birds, it was hypothesised that an index of habitat quality could be built using the remotely sensed data. Since it is known that wet sites would produce low band 7 reflectance and dryer, more productive ones would produce higher values, it was further hypothesised that there should be an *inverse* relationship between the mean infra-red reflectance value for the site and the numbers of wading birds present.

The model was tested using data for bird counts collected from transect surveys of 42 moorland sites, each of 6.4 km^2, collected in 1986. A linear regression model was constructed, relating population numbers at a site to the corresponding mean band 7 reflectance value derived from the MSS image. The most successful model developed was for dunlin, where about 60% of the variation in numbers was explained (Table 19.1).

The dunlin model was used to predict numbers throughout the study area for land which had not been assigned to one of the 'unsuitable' cover types (i.e. open water, forestry, farmland, etc.). These data could then be used to assess the area of habitat in various quality classes and the extent of the losses due to afforestation by overlaying the forest stock information for 1988.

In order to test the results of the study critically, the map output was used to make population predictions for a set of sites selected at random within the area. A second phase of fieldwork was then undertaken to compare these estimates with

Table 19.1. Linear regression models for waders based on band 7 (B7) reflectance (1987 study).

Species	r	N
Dunlin	−0.781	15.23−0.215 × B7
Greenshank	−0.425	2.36−0.030 × B7
Golden plover	−0.643	11.77−0.156 × B7

r, Linear regression coefficient; N, wader count at site; B7, mean MSS band 7 reflectance value for site.

those obtained on the ground. It was found that there was no significant difference between the numbers observed and the numbers predicted.

On the basis of these initial results it was felt that some assessment of the impact of forestry in the Flow Country could be made. Between 1978 and 1988, 18% of the Flow Country was lost to afforestation and we predict dunlin populations have declined by 17%. These data compare with independent estimates (Bainbridge *et al.*, 1987) of about 20% decline for the same general period.

Testing the landscape model

Despite the success of the initial phase of the study a number of questions remained. One area of concern focused on the extent to which the remotely sensed data were capable of mapping the distribution of the dubh lochan systems. Standing water, whether in the form of lochs or dubh lochan, may absorb infra-red strongly, but these habitat features exist only as patches in an extensive vegetated matrix. For water to play a systematic role in suppressing the infra-red signature of an area, the reflectance of the matrix must remain nearly constant between areas. This assumption was tested critically in a second phase study undertaken in 1990.

Figure 19.1 shows the extent to which pixels in different infra-red reflectance categories may suppress the overall mean. The range of infra-red values which push the mean reflectance up or down for a given area are in the central part of the distribution. Even though each individual pixel in this region departs only slightly from the all-site mean, the sheer number of these matrix pixels means that the influence is a strong one.

The reflectance characteristics of the matrix vegetation may, however, still be relatively constant if sites contain a random mix of pixels from around the mean. Two examples of frequency histograms for individual sites (Figures 19.2 and 19.3) show that the modal position of the matrix pixels on the infra-red scale can vary considerably. Figure 19.2 shows the site with the highest infrared reflectance sur-

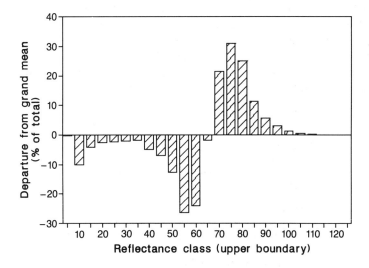

Figure 19.1. Potential of pixels in different band 7 classes to influence the mean reflectance of a given area.

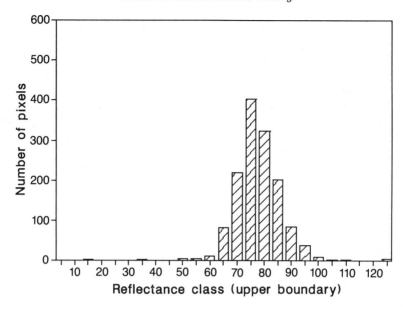

Figure 19.2. Frequency histogram of pixels for a site with mean band 7 reflectance of 75.6.

veyed in 1986, it has no pool systems and two small lochs. Figure 19.3 shows data for a site visited in 1990; apart from two small pool systems and some small lochs, which produce the slightly drawn out left-hand tail of the figure, the site is virtually a *Tricophorum caespitosum* monoculture. The mean infra-red reflectances for the 1986 survey sites range from 53.0 to 75.6. The means for the two sites in Figures 19.2 and 19.3 are 60.9 and 75.6, respectively. They span 65% of the total variation across sites. The difference between them is almost entirely due to vegetation char-

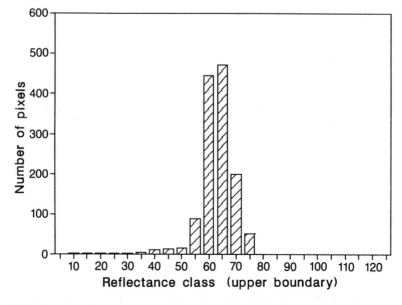

Figure 19.3. Frequency histogram of pixels for a site with mean band 7 reflectance of 60.89.

acteristics. These data suggest that the assumption about the constancy of the spectral reflectance of the vegetation matrix must be rejected.

In contrast to the matrix, lochs (returning pixels values between 0 and 25) have less total effect; the scarcity of lochs in relation to the matrix overcomes the very low infrared reflectance of each individual pixel. Dubh lochan systems, being part water, part vegetation, have spectral signatures between that of lochs and vegetation, and have only a local influence on infra-red reflectance (Figure 19.1). However, despite the fact that the matrix tends to play a significant role in controlling the infra-red reflectance of a given area, all three habitat types—matrix, lochs and dubh lochan systems—tend to pull in the same direction: areas of low matrix reflectance correspond to the bare *Tricophorum* dominated blanket peat in which lochs and pool systems tend to be more common.

Refining the landscape model

Because the wader/infra-red correlation is more complex than originally thought a new habitat model was developed. Since dunlin and golden plover rely to some extent on dubh lochan systems and the spectral signature of such patches is distinctive, it was thought that the proportion of such habitat at a site may be a better predictor of numbers than the overall site mean. The 1986 survey data were reanalysed to determine the range of infra-red values which gave the best prediction of wader numbers, when the numbers of pixels in that class were expressed as a proportion of the total number of pixels for the site. In this way, the best predictor range of band 7 values could be identified regardless of its position. Moreover, the prediction that it should correspond to the infra-red domain of pool systems could be tested.

Figures 19.4 and 19.5 show the results of the analysis for dunlin and golden plover using non-parametric methods (Spearman rank correlation). Small intervals of the band 7 range were chosen for this analysis to highlight the trend of correlation. For the predictive model using linear regression, a wider band of ten was

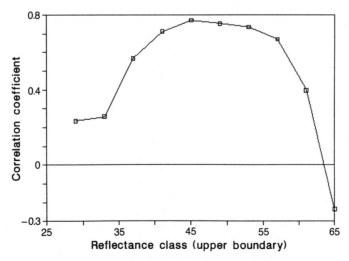

Figure 19.4. Correlation of dunlin numbers with the proportion of habitat at a site in different band 7 ranges.

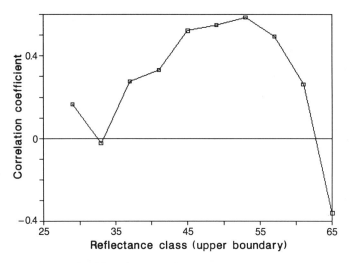

Figure 19.5. Correlation of golden plover numbers with the proportion of habitat at a site in different band 7 ranges.

used. The best three predictor ranges from the non-parametric correlation were chosen (giving a band of twelve DN values) and the middle range of 10 was used for building the regression model. This procedure gave the same predictor range for both dunlin and golden plover, i.e. band 7 values between 43 and 52. The final site quality index was calculated as:

$$\text{SQR (No. pixels in the range 43–52/Total No. pixels)}$$

The square root (SQR) transformation was required to make the relationship of the index with the 1986 bird counts linear. For dunlin the procedure was a disappoint-ment; the correlation is only slightly better than the results of the initial study (0.8 as opposed to 0.78). The prediction of bird numbers for the 1988 survey is only slightly better than that obtained using the site mean (predicted–observed corre-lation of 0.73 as opposed to 0.68). The main improvement obtained using this new model of landscape structure was in the better correlation obtained for golden plover. The correlation coefficient increase from 0.64 to 0.73, giving a 19.3% increase in explained variance.

For the 1988 survey data alone, the habitat index also performs better for both birds than the site mean. For dunlin the improvement is from 0.68 to 0.73, and for golden plover from 0.71 to 0.76.

It is interesting to note, however, that neither the mean nor the habitat index seems capable of predicting the number of golden plover at a site for the 1988 survey data (Figure 19.6). The scatter of points is quite large, but the slope also seems to be quite wrong, approaching 1.0 instead of the predicted 0.4 (taking into account the difference in survey effort in 1988).

Bird distribution and landscape structure

Although the same range of pixel values gave the best predictor for both dunlin and golden plover, this does not imply that the birds have the same habitat preference

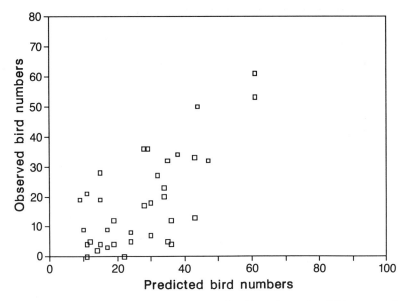

Figure 19.6. Predicted versus observed numbers of golden plover for 1988 survey. Prediction based on habitat index (see text).

or spatial distribution at a finer spatial scale. In fact field observations suggest that, although dubh lochan patches may be the most important habitat feature for both birds, golden plover are more catholic in their tastes. To investigate this, the within-site affiliation of each species to habitat with different spectral signatures was assessed.

The algorithm used in this phase of the study consisted of a routine which transformed the Ordnance Survey grid coordinates for each bird sighting ($n = 892$ for dunlin, $n = 987$ for golden plover) into corresponding image coordinates. It then searched the image around each bird until a pixel was found within a specified band 7 target range. The distance between each bird and the nearest target pixel was recorded. The same procedure was also carried out for a set of random coordinates, the same number of random coordinates at each site as there were birds of each species.

Figure 19.7 shows how dunlin (lower curve) tend to cluster more tightly around areas of lower value pixels than golden plover. Both dunlin and golden plover were closer to areas of low reflectance than was available at random. The preference is most marked for dunlin.

The observation that dunlin and golden plover are distributed differently at finer spatial scales has allowed the production of habitat suitability maps for the two birds based on the 1986 bird counts.

The procedure adopted was as follows: the band 7 value of each pixel occupied by each individual bird was recorded. The data were grouped to produce the number of birds in each band 7 category (categories as for the non-parametric correlation above). The total area of habitat available in each category for the 1986 survey sites was calculated in the same way. The number of birds occupying a particular habitat category divided by its availability gives the density of birds (birds per 1 km × 1 km square) one would expect to find in habitat of that type, given a similar survey method.

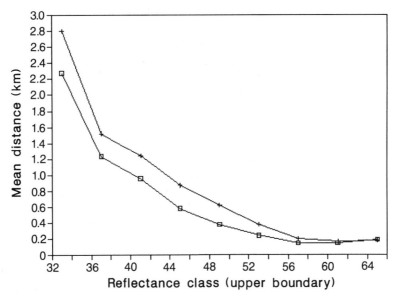

Figure 19.7. Mean distance (km) between pixels in different band 7 ranges and dunlin sightings
(□), golden plover (×).

The actual densities for dunlin and golden plover are shown in Figure 19.8.
The first four ranges have been deleted as it is known that pixels in this range
correspond to areas of continuous open water, and neither bird swims. The upper
four ranges have also been deleted for both birds because the low number of birds
and the rarity of habitat in these ranges means that chance factors can make the
predicted density calculation behave wildly. The deletion of these outer eight ranges
leaves 96.2% of dunlin and 98.2% of golden plover in the analysis.

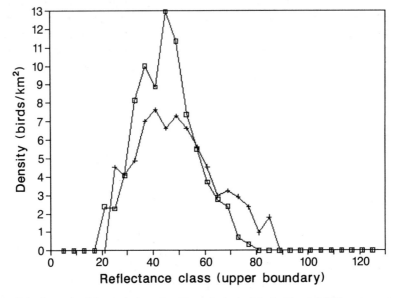

Figure 19.8. Density of dunlin (□) and golden plover (×) in habitat of different spectral reflec-
tance.

As expected, the distribution is slightly wider for golden plover than for dunlin (Figure 19.8). Also, where the habitat is most suitable, dunlin attain higher densities. Mapping this information it becomes clear that dunlin are predicted to reach far higher densities than golden plover where habitat is suitable. However golden plover, while never reaching densities attained by dunlin, tend to maintain their density at higher levels where habitat is less suitable.

Conclusions

Although the models of landscape structure have been refined and better predictions have been achieved for golden plover, many questions remain. In particular we need to consider how sensitive the models are to the characteristics of the image data. Further work is required in order to test the robustness of the models to imagery derived from other sensors for other seasons and other years, and to explore whether multi-date imagery can improve the level of explanation achieved by the existing habitat models.

We also need to test over what geographical area the models are effective in predicting variations in the numbers of wading birds. Further work is required to determine whether the models derived from the Flow Country can be applied to other areas of similar habitat which are known to support significant wading bird communities, such as the Shetland Islands. This will enable a deeper understanding of the relationship between near-infrared reflectance and bird numbers to be achieved.

Finally, we need to investigate the temporal changes in pool systems themselves. Maintenance of these key habitat patches is vital in preserving the ecological value of the Flow Country for wading birds. The stability of such habitat features needs to be investigated if the models of landscape structure developed here are to be fully corroborated.

Despite these questions, however, it does seem possible to use remotely-sensed image data to model spatial variations in populations of several key wading bird species. The models depend on the extent to which image data can detect and quantify landscape structure within the Flow Country. As the pressure of human activities on the biosphere increases, we need to develop such tools for predicting environmental impact and assessing the consequences of our policy goals. The Flow Country represents a useful 'natural laboratory' in which techniques can be developed.

Editor's Note

A complete account of this work is available from the authors.

References

Avery, M. I. and Haines-Young, R. H., 1990, Population estimates for the dunlin *Calidris alpina* derived from remotely-sensed satellite imagery of the Flow Country of northern Scotland, *Nature*, **344**, 860–862.

Bainbridge, I. P., Housden, S. D., Minns, D. and Lance, A. N., 1987, *Forestry in the Flows of Caithness and Sutherland (Conservation Topic Paper No. 18)*, Royal Society for the Protection of Birds.

Index